Fundamentals of

Biochemical
Calculations

Second Edition

Fundamentals of
Biochemical
Calculations

Second Edition

Krish Moorthy

CRC Press
Taylor & Francis Group
Boca Raton   London   New York

CRC Press is an imprint of the
Taylor & Francis Group, an informa business

CRC Press
Taylor & Francis Group
6000 Broken Sound Parkway NW, Suite 300
Boca Raton, FL 33487-2742

© 2008 by Taylor & Francis Group, LLC
CRC Press is an imprint of Taylor & Francis Group, an Informa business

International Standard Book Number-13: 978-1-4200-5357-9 (Softcover)

Library of Congress Cataloging-in-Publication Data

Moorthy, Krish.
　　Fundamentals of biochemical calculations / Krish Moorthy. -- 2nd ed.
　　　　p. ; cm.
　　"A CRC title."
　　Includes bibliographical references and index.
　　ISBN-13: 978-1-4200-5357-9 (hardcover : alk. paper)
　　ISBN-10: 1-4200-5357-4 (hardcover : alk. paper)
　　1. Biochemistry--Problems, exercises, etc. 2. Biochemistry--Mathematics. I. Title.
　　[DNLM: 1. Biochemistry--Problems and Exercises. 2. Mathematics--Problems and Exercises. QU 18.2 M825f 2008]

　　QD415.3.M66 2008
　　572.01'51--dc22

　　　　　　　　　　　　　　　　　　　　　　　　　　　　　　　　　2007020734

Visit the Taylor & Francis Web site at
http://www.taylorandfrancis.com

and the CRC Press Web site at
http://www.crcpress.com

To teachers and students of biochemistry

Contents

Preface to the First Edition

Biochemistry, like chemistry, is a quantitative science. Even though it may be taught in a descriptive manner, the research involved in establishing the descriptive principles, a number of principles themselves, and in today's application of biochemistry to molecular biology, applied biology and biotechnology, the quantitative aspects and approach are important and integral parts of biochemistry. Yet, many students at all levels experience difficulties with biochemical calculations. One reason for this problem may be that students enter certain courses with the notion that they have chosen a biological (descriptive) stream of studies and see the quantitative aspects of biochemistry as something foreign to their expectations. Another is the general decline in numeracy skills in students entering tertiary studies even in the more mathematical streams, such as engineering and physical sciences—a concern expressed by many educators. During my experience with tertiary students, I have tried to identify the reasons why students experience difficulties with chemical/biochemical calculations. Some of the reasons, I believe, relate to over-dependence on memorised formulae, dependence on memorised definitions, and in "formulae-juggling," all of which result in a lack of feel and the inability to reason logically. This is further discussed in Chapter 1. It is hoped that students may be able to identify their individual problems (if any) and set about remedying them.

Fundamentals of Biochemical Calculations recommends the use of the ratio method in solving biochemical calculations because it is simple, logical, commonsense-based, requires no formula to be memorised and one that provides considerable feel. This approach depends on earlier experiences in chemical calculations. Accordingly, most chapters, particularly Chapters 2 and 3, start with high school or first year chemical calculations and lead the student into biochemical calculations. It is for this reason the word "Fundamentals" is included in the title—this book does not include the more advanced or special technique-oriented calculations. Likewise, no advanced knowledge of mathematics is required beyond the elementary logic behind mathematical operations acquired in junior secondary school. This book is directed at students in their first one or two years of biochemical studies but may also prove useful to laboratory technicians and postgraduate students who are non-majors in biochemistry working in any biological discipline. At the start of each chapter is an introductory discussion. It is hoped that the student will find these discussions challenging; the purpose of these once again is to revise and recall earlier experiences and to see if the student would agree with the logic behind the problems. Important comments relating to practical understanding and situations are also included in these introductions. It is in the practical situations that biochemical calculations have their especial applications. All problems are completely solved, showing detailed calculation steps and including, where appropriate, comments and mathematical hints. The intention is to re-enforce mathematical logic and to provide foundations for the development of reasoning skills. Fully worked-out solutions are designed to show (as opposed to teach or instruct) how to perform biochemical calculations.

I have been fortunate enough to have taught biochemistry to over 2 000 students in more than 25 years as lecturer in various disciplines—applied biology, biotechnology, clinical sciences, food technology, toxicology, medical radiation sciences, and physiotherapy. I am grateful to my students for their interactions with me and in particular for the feedback on the solutions to the calculation questions presented over the years as Problem Sheets. It is their response that prompted me to write this book and submit it for general distribution.

I wish to express my appreciation to two of my colleagues, Dr. Theo Macrides and Dr. Dave Propert, for their advice and review of the original manuscript, and to Professor Andy Sinclair for his support and encouragement.

This book is dedicated to my daughter, Lisa, and my wife, Maria. Lisa, now a graduate in biochemistry, for providing me one further avenue for expressing my love—teaching her biochemical calculations, and Maria for her unending love, support, patience, and faith always and especially during the writing of this book.

Krish Moorthy
Royal Melbourne Institute of Technology
March, 1994

Preface to the Second Edition

The First Edition of this book was an author-published, limited edition used in my teaching at RMIT-University, Melbourne, Australia, and distributed within Australia. It was well reviewed, the most comprehensive review being the official review by John Swann and Tony Dawson for the Australian Society for Biochemistry and Molecular Biology (1994). In that review, they concluded: "We feel that Krish Moorthy's book should find a useful place on the bookshelves of many teachers and students of biochemistry."

Over the last few years, as copies of the First Edition ran out, some of my colleagues have complained about the unavailability of the book for prescription in their courses. It was the good reviews, the requests from my colleagues, and the never-ending encouragement from my wife that prompted my attempt to bring out this, the Second Edition.

While staying close to the original purpose and scope of the book, this edition has been completely revised and updated. Advice on how to gain confidence over biochemical calculations has been extended throughout the book and additional *questions have been added to all chapters. Three new chapters have also been added:* "Practical Calculations," dealing specifically with data handling, calculations, and presentation of results; "DNA and Molecular Biology," to reflect the emerging techniques; and "Pharmaceutical Calculations," to both extend the scope of this book and to reflect the fact that the ratio method of calculations, promoted in this book, is also recommended by pharmacy teachers.

I am grateful for the helpful suggestions and comments from lecturers and reviewers from Australia, the UK, and the States. Some of their ideas have been incorporated into this edition. However, as the scope of this book is to introduce *fundamental (mathematical)* techniques for performing biochemical calculations, the more advanced biochemical techniques and the mathematics that go with them are not included here. Accordingly, this book does not contain calculations relating to GLC, HPLC, and MS traces and there are no computer-based analyses or spreadsheets.

My deep gratitude also goes to Lindsey Hofmeister, Commissioning Editor, Chemistry, Taylor & Francis, CRC Press, for her advice and encouragement and indeed commissioning this edition. It is with great pride, yet humility, I follow the words of John Swann and Tony Dawson in dedicating this edition to "teachers and students of biochemistry" throughout the world.

Krish Moorthy
Melbourne, Australia

Supplemental Material

A test bank of supplementary questions (with answers) accompanying this edition is also available. Please contact Susie Carlisle, Academic Sales Manager, at susie.carlisle@taylorandfrancis.com to receive a copy of this CD.

1
Introduction

Biochemistry students have generally studied the basic quantitative aspects of the subject in earlier introductory chemistry subjects that have elementary mathematical concepts as a prerequisite. However, even though they may have passed chemistry and mathematics, and possibly even physics, many students experience difficulties with biochemical calculations. There may be a whole host of reasons for this problem. The reader is invited to explore some of the possible causes.

1. Over-reliance on Formulae

In mathematics and quantitative sciences, almost every principle or situation has been neatly packaged into a formula. Thus we have:

$$E = mc^2$$
$$D = \frac{M}{V}$$
$$PV = nRT$$

$$\text{surface area of sphere} = 4\pi r^2$$

$$\text{sum of the interior angles of a polygon} = 2n - 4 \text{ right angles.}$$

Unlike 40 or 50 years ago, less time seems to be spent on the derivation of the formula. This virtually robs the student of any feel for what is going on. Instead, the bulk of teaching these days seems to focus on the application of the formula. Students memorise the formula and substitute values (which in high school examples are often more easily substituted) and proceed with the calculation. The formula takes the all-important role and the basic understanding is often lost. So much so that, for example, two years later, if students were asked, "What is density?", the response could be "D over V" or they might correct themselves by saying "no, it's M on V." Seldom would they even mention the words *volume* or *mass*. This is guesswork, and it is this sort of approach that causes problems for students of biochemistry. The expected and proper response is "density is the mass of unit volume (e.g., cubic centimetre) of a substance." This response is a positive indication of understanding. Students who respond in this manner will confidently divide the mass by the volume, knowingly and feelingly, to obtain density.

Let us consider another example: PV = nRT. Perhaps this formula made perfect sense to you two or three years ago or you remember applying this formula and passing your exams. How about now? Two years later or twenty years later if you were asked, "What happens to a volume of gas (say 1.2 L) when the temperature is increased from 20°C to 40°C?" If your gas laws were simply thrust upon you and you used them without basic understanding or feel:

(a) You would not have the faintest idea how to solve the problem.
(b) You would try very hard to recall the formula.
(c) If you do recall the formula, you would want to be given the value of R.
(d) You might not know which temperature scale to use.

On the other hand, had you understood what was involved in the derivation of the formula:

(a) You would reason out that volume increases (the gas expands) as temperature rises.
(b) When you start thinking along these lines, the mental stimulus would start you thinking about matter expanding with absolute zero as the baseline (and therefore dismiss the Celsius scale because it is only an arbitrary scale relating to water).
(c) You might even recall the number 273 and figure out that −273°C is absolute zero and that 20°C = 293 K, and 40°C = 313 K.
(d) Because you "felt" that gas should expand, you would set up a calculation that would generate an answer greater than 1.2 L:

$$1.2 \times \frac{313}{293} \text{ L}$$

This gives us exactly the right answer (and PV = nRT was not needed!).

This scenario exemplifies the sort of confidence, common sense, logic and feel that is expected of a mature student doing biochemical calculations. Logic and common sense can be used with confidence if you can derive relationships from first principles. Dependence on memorised formulae might have been all right for short periods, for substituting fairly directly given values, for passing exams the next day or the next month, but how will that serve you years later and in new situations?

2. Use of the Factor-Label Method in Calculations

In your earlier years in physics and chemistry, you might have come across the factor-label method for performing calculations. This method advocates that in calculations involving physical quantities, the units should be multiplied or cancelled as if they were numbers. In other words, use the formula method but make sure to carefully enter all the units (that describe the physical quantities) alongside the numbers and then multiply or cancel them.

If the unit that remains correctly describes what you were after in the first place, then you must have used the formula correctly!

Let's try an example using the equation PV = nRT. Say we want n in a particular question, so:

$$n = \frac{PV}{RT}$$

Let's play this game further, let's say P = 13 000 Pa, V = 0.02 m³, R = 8.31 m³ Pa K⁻¹ mol⁻¹, and T = 300 K. So:

$$n = \frac{13\,000 \text{ Pa} \times 0.02 \text{ m}^3}{8.31 \text{ m}^3 \text{ Pa K}^{-1} \text{ mol}^{-1} \times 300 \text{ K}}$$

$$n = \frac{13\,000 \cancel{\text{Pa}} \times 0.02 \cancel{\text{m}^3}}{8.31 \cancel{\text{m}^3} \cancel{\text{Pa}} \cancel{\text{K}^{-1}} \text{mol}^{-1} 300 \cancel{\text{K}}}$$

n = (something) mol

What do you know! You are left with n = mol (the number of moles), so everything in the equation must have been entered correctly. But did you gain any understanding about what was really happening scientifically? (Using a formula does not exactly create a mountain of understanding or feel; and then when you go and strike off the units within it, all you are doing is picture-editing.)

Let us examine another case of "formula juggling"; this one is in relation to an answer to a physics question. (While, it is acknowledged that neither example deals with biochemical problems, they are the very questions — and method of solution — that students grow up on. Biochemistry students often try to adopt the quantitative skills they learned in their previous year's physics and chemistry courses for solving biochemical calculations. They would, however, be far better off handling biochemical calculations if they depended more on their primary school mathematical skills than on what they learned in university physics or chemistry.)

The Physics Question: An outdoor loudspeaker system placed on a high tower gives out sound of equal intensity in all directions. To test this system, a sound of steady frequency and amplitude is fed through it, and sound meters, placed 10 m from the speaker system, are used to measure the intensity. The intensity is found to be 3.0×10^{-6} watts m^{-2} and is the same in all directions. Calculate the *total acoustic power in watts* produced by the speaker system. Given: surface area of a sphere = $4\pi r^2$.

Using the factor-method, the solution would look like this:

$$I \text{ (intensity)} = 3.0 \times 10^{-6} \text{ watts m}^{-2} \tag{i}$$

$$\text{Area of sphere} = 4\pi r^2$$

$$= 4\pi (10)^2 \text{ m}^2 \tag{ii}$$

Power is required in watts. Using the clue of units, we need to multiply (i) by (ii):

$$P = 3.0 \times 10^{-6} \text{ watts m}^{-2} \times 4\pi (10)^2 \text{ m}^2$$

(m^{-2} cancels with m^2, leaving behind just watts)

$$= \mathbf{3.8 \times 10^{-3} \textbf{ watts.}}$$

The basic points in the calculation are:

1. The average intensity of the loudspeaker is calculated as **watts m^{-2}** directed at one square metre of surface area of an imaginary sphere.
2. The surface area of the sphere, naturally, comes out as **m^2**.

Simple logic would tell us that having obtained the intensity for just one square metre, we need to multiply this value by the total surface area in order to get the intensity for the *whole* surface area. Instead, what this approach is suggesting is: "Using the clue of *units*, we need to multiply." That is, because one part of the answer comes out as **watts m^{-2}** and the other as **m^2**, to get **watts**, we must multiply (to get the answer in the correct units, which in this case is **watts**).

The emphasis is not on scientific reasoning but simply on obtaining the right units. If the required formula is precisely entered (along with the units for each term!) and the factor-label method is strictly adhered to, then the correct answer will be produced. In other words, if you juggle the formula into place and continue to juggle around the units, you will juggle yourself into the correct answer. It does work! The focus is all on the formula and the units, and any logic-statement that is made is in reference to getting the units right. This approach might provide the right answer, but it does not provide any basis for building scientific foundations and understanding.

3. The Dilution Factor Method

The dilution factor method is a calculation method used by some students. Consider the following examples:

1. 2.0 mL mol/L solution is diluted to 200 mL. The dilution factor (DF) is 100.
2. 0.2 mL is diluted to 2.0 mL. The DF is 10.
3. 0.01 mL is diluted to 25 mL. The DF is 250.
4. 1.76 mL of an extract is diluted to a final volume of 400 mL. The DF is $\frac{400}{1.76} = 227.272727...$ = 227.27.
5. 0.3 g is taken from a total mass of 46 g. The DF is $\frac{46}{0.3} = 153.3333... = 153.33$.
6. 46 mL ethanol is mixed with 92 mL water. The DF is 2.

It is not necessary to obtain dilution factors separately in calculations. There are two reasons why this practice is discouraged:

(a) There is a temptation to do simple calculations for the dilutions in the head. Mistakes can be made when doing these steps mentally. Furthermore, as the original figures are not written down and used with other steps in the overall calculations, it is not possible to check if correct figures have indeed been used. In the ratio method recommended in this book, you are required to actually write down $\frac{200}{2.0}$ or $\frac{2.0}{0.2}$ or $\frac{25}{0.01}$. (Incidentally, the last expression works out as 2500 and not 250, as shown in Example 3.) The figures will take care of themselves when the other steps in the calculations are included and computed, and there might be some convenient cancellations with other entries in the overall calculations. In some calculations (e.g., if you were asked for the original concentration), the "dilution factor" of 100 may be misleading and indeed the expression might have to appear as $\frac{2.0}{200}$, which is 0.01, not 100. The adoption of the ratio method (to be described shortly), on the other hand, provides a better feel for the appropriate steps to be taken, including the expression of the ratios in the correct orientation.
(b) When complex figures are involved, as in Examples 4 and 5, the figures have to be written down and dilution factor calculated (or done on an electronic calculator). Performing such calculations is a waste of time and, if the answer obtained is rounded-off, round-off errors could be introduced. The ratio method requires that you leave $\frac{400}{1.76}$ and $\frac{46}{0.3}$ as they are, incorporate them with other calculations, and round-off (to appropriate number of significant figures) only once at the end to obtain the final answer.

Also, consider the following:

(c) The term "dilution factor" really does not make sense when dilutions are not involved (Example 5). Taking 0.3 g out of 46 g is not a dilution.
(d) When the ratio method of calculation is used, a separate so-called "dilution factor" is not necessary.
(e) The dilution factor in example 6 is not correct. It should be $\frac{92+46}{46}$, which is 3 if one assumes that the volumes are additive. (Ethanol actually dissolves in water and, therefore, the final volume is less than 92 + 46, or 138 mL. This volume must be known to work out its "dilution.")

4. The Guided-Tour Method

The guided-tour method is an approach taken by instructors who have no time to show students how to perform (and understand) calculations in chemistry, physics, or biochemistry. It often rears its head towards the end of practical sessions. For example, you have obtained some data from your experiments. You need to do some calculations in order to get the final *results*. However, you do not know how to calculate, or you might take too long (and it's already lunchtime) or you might make mistakes, so a dummy run of the calculation is provided for you on the Technical Report Sheet or on the board. All you have to do is plug in your values and perform one or two multiplications — and there you have it. Not a bad idea at all; you can proceed to look at the *scientific* implications of the results. But when are *you* going to learn how to calculate?

Courses for biochemistry majors do expect students to be able to perform the whole of the calculations for themselves. In postgraduate courses and research, there are no guided tours.

5. The Mole Concept

The concept of the *mole* is central to virtually all aspects of quantitative chemistry and biochemistry. Yet, science students even at second-year level do not seem to have a full appreciation of this fundamental concept.

Once again, about 50 years ago, students had fewer problems. At that time, the *mole* was called *gram-molecular weight*. It was accepted as an amount in grams of a substance. The renaming to "mole" is not the problem. What seems to have caused the problem is the definition that it is a number. Definitions, like formulae, have become beautifully and neatly packaged and the student's reliance on them is disturbing. At tertiary level, students cannot hide behind definitions or formulae. They have to fend for themselves with basic, raw understanding. The mole concept is further discussed in Chapter 2.

6. Units as Exponentials

This is the practice of writing combined units in power terms (e.g., density as g cm^{-3} and acceleration as m sec^{-2}). This habit started in physics and has now moved on to chemistry and biochemistry. Clearly, this form of expression is related to formula usage and the factor-label method of calculations. Using this method, density, which is *g per cubic centimetre*, or *g/cm^3*, becomes g cm^{-3}. When two or three similar units appear in combined formulae, instead of appearing as denominator and numerator and then cancelling, the positive and negative powers are cancelled (actually summed); for example, sec^{-2} sec^{+1} = sec^{-1}.

There is nothing whatsoever wrong with this. The problem is that biochemistry students do not gain as much feel for these cryptic expressions as they do for the use of the word "per," the slash (/) symbol, or words such as "square" or "cubic." This reduced feel affects their understanding. (Other aspects of exponentials are discussed in Chapter 2.)

7. Scientific Notation

Some students coming from physics and chemistry backgrounds, thinking that scientific values must be written in scientific notation, and proceed to write 1×10^{-1} instead of 0.1 and write 2.34×10^2 instead of 234. The problem for biochemistry students is that the scientific notation of writing number for values such as chemical amounts or volumes does not create as much feel as using prefixes (coming up in the next chapter). For example, *10 cents* creates more feel than *$0.10*. (This is because we immediately get the feeing that we are at the low level of currency, whereas *$0.10* is just symbolic.) Likewise, using 10 mmol/L creates more feel to a biochemistry student (who is still learning basic concepts of concentrations of body fluids) than 10^{-2} mol/L. Again, please see Chapter 2 for further comments.

The student would have noticed that in the foregoing discussion one particular point kept repeating itself: adopt methods that provide more feel. Doing something with feeling is understanding.

All knowledge is based on understanding. Gaining the knowledge to perform biochemical calculations is no exception.

The Ratio Method

What then is the best method to be used? This book recommends using the ***ratio method*** as much as possible for solving biochemical problems.

Before going into the details and advantages of the ratio method, let us recap who would be performing these calculations and under what circumstances. The student is most likely to be in his or her second year. It is a year or two since he or she has performed formal and routine chemical calculations. This student is not expected to immediately recall the various chemical (calculation) formulae for performing chemical calculations. The student however is relatively mature and has confidence in his or her basic thinking and reasoning skills. This student trusts himself or herself in making simple logical decisions on mathematical concepts (such as multiplying by a number greater than one will increase the value, or dividing by a number less than one will also increase the value). This student will not blindly try to recall a formula and seek shelter from it. This student will not try juggling things about in the hope the answer might turn out right.

Such a student will systematically employ the ratio method, which simply involves dealing with ratios or proportions. Each extra step will take an extra fraction as a multiplicand. See answers to Questions 1 and 2 in Chapter 11 to gain a full impact of this method, but for the moment consider this:

Question: If 8 bananas cost 70 cents, what is the cost of 5 bananas?
Answer: The ratio method requires exactly this set-up for the calculation:

$$8 \text{ bananas cost } 70 \text{ cents}$$

$$\Rightarrow 5 \text{ bananas cost } \frac{5}{8} \times 70 = \textbf{43.75 cents}$$

It is important that the expression be $\frac{5}{8} \times 70$, not $\frac{70}{8} \times 5$, and that the bananas be kept together in one "block" and the cents be kept in another "block." The basis of the ratio method is that the 70 cents becomes a *smaller* value by the ratio of $\frac{5}{8}$.

Question: If 8 bananas cost 70 cents and 5 children are allowed to eat $2 worth of bananas, how many bananas will be eaten by 21 children? (*What a crazy question, but bear with me.*)
Answer: 70 cents is the cost of 8 bananas

$$\Rightarrow 200 \text{ cents is the cost of } \frac{200}{70} \times 8 \text{ bananas}$$

(the $\frac{200}{70}$ makes the answer *bigger*)

$$5 \text{ children eat } \frac{200}{70} \times 8 \text{ bananas}$$

$$\Rightarrow 21 \text{ children eat } \frac{21}{5} \times \frac{200}{70} \times 8 \text{ bananas}$$

$$= 96 \text{ bananas}$$

(the $\frac{21}{5}$, makes the answer *bigger*)

Note the children "block," the money "block," and the banana "block" as well as the build-up of a *string of blocks* as additional steps are introduced. We will call these fractions *Multiplicand Ratios.* Note the effect these multiplicand ratios have on the overall value. This is the ratio method — virtually all the questions in the book can be solved using this method. No further comments will be made at this stage, except to draw your attention to the fact that this method intrinsically involves an important but simple decision-making step — the orientation of the multiplicand ratio. This, in turn, means *you* are shaping, therefore evaluating, every step of the calculation. (The students I have taught have dubbed the ratio method "the banana method.")

The setting out, too, is very important. Make certain that what has to be solved appears in the second line on the right-hand side.

$$A \text{ is to } B$$
$$As \; C \text{ is to } \frac{C}{A} \times B$$
$$\uparrow here$$

And, there is more. With $\frac{C}{A}$ you cannot feel if the overall answer is getting bigger or smaller. *With real numbers you can, and in the ratio method, you are supposed to.* Truly, all of this is primary school stuff!

Please note, while this book is advocating the use of the ratio method, the more important message, is that students use methods that they are familiar with and feel confident about. If you are doing a particular set of calculations and are constantly using a formula, or can recall a formula or have access to it, or are given the formula, by all means, plug the values into the formula and proceed. Just be sure to evaluate your final answer.

Electronic Calculators

Without a doubt, electronic calculators are marvelous inventions. Make sure you are totally familiar with your model. (It is a poor show if you have to borrow an unfamiliar one in the exam room.) Note that the ratio method is ideally suited for computation with the aid of an electronic calculator. Original values (exactly as given in the question) are set out as a *string of blocks;* each entry on the top line is multiplied, and then each bottom line entry is divided one at a time. *Repeat the entire calculation once more and check that you get the same answer.*

Because electronic calculators save incredible amounts of calculation time, use the time saved in checking that correct figures are copied from the question, gaining a good feel of the question, appreciating the steps in your calculation, and evaluating your answer by whatever means possible. One of the common mistakes that students make with electronically calculated answers is having the right figures but with a wrongly positioned decimal point. (Mistakes commonly occur when entering negative exponents. Remember to hit the +/− key and then pause to check that the "−" indeed appears. Another very common mistake is entering 3×10^{-3} as **3, ×10, exp-3** when it should be **3, exp-3**.) Make sure your answers are sensible, and present your final answers with the appropriate number of significant figures.

Note, the following are nonsensical answers:

28.46 g glucose/mL

146.17 mL diluted to 100 mL

$$1.4617 \times 10^{-12}\,\text{mL diluted to } 100\,\text{mL}$$

$$227.27272727\,\text{mol/L}$$

$$8.07609743\,\text{g}$$

How to Use This Book

1. Read the introductory sections of each chapter. The material is presented in a challenging manner. See what you think about the comments. Do you agree? (The theory included in the introductory sections is designed to provide only a brief background to the questions covered in that chapter. In some chapters, you may be required to use the values given in the introductory section for solving the questions that follow.)

2. Attempt all questions in the order they are given (if not otherwise instructed by your lecturer or tutor). Check each of your final answers. If your answer matches the answer given and you have performed the steps in your calculation confidently, then do not concern yourself too much about the method given in this book. Have a look through it, though. The method given in the book might be shorter, simpler, or more logical; avoid dependence on memorised formulae; avoid confusion and risk of errors (including round-off errors); enable better evaluation of the answer; or contain hints that will provide confidence for solving other problems.

3. The answers to the questions are not placed immediately below the questions. The intention is to avoid your looking at the answers and thinking "Oh, yes, that looks simple enough; I can do that." You are meant to try each question without any hints or clues ("guided tours"). Try to develop the clues yourself. It is really quite simple if you give yourself time. Ask yourself:
   - What is the question really asking and in what units must I give the answer?
   - What information is given?
   - How do I go from that to that — from the information (in its units) to the answer (in the units required)?

   Finally, set out your working tidily (see item 7). You will not be able to stop the answer from falling into place. And, you will be able *to see* that it should be the correct answer because *you* have guided its falling into place all the way.

4. While pondering the question following the above guidelines, use a pencil or pen to underline the key figures in the question before proceeding.

5. If you have to consider the fully solved answer, go through each step very carefully (again use a pencil or a pen). Do not simply adopt the view that you recognise what is happening or that you agree with any step. See that you understand and gain a feel for what is set out. Carefully note any comments provided. Please go through the steps at least twice. It is better to write down the steps to gain a better feel rather than merely looking at them.

6. Go back a few days later and reattempt the questions that you had difficulties with.

7. Tidy, compact writing is a very important asset for mathematicians and quantitative biochemists. It is never too late to improve your writing and presentation. If you set out your work properly, you will see things better. Have a close look at the presentation of the solutions in this book, in particular the stringing of blocks of multiplicand ratios mentioned earlier, and the lining up of equal signs and mathematical expressions.

8. You can create further questions by doubling or halving some value in the question. Predict how the answer should change. Reattempt the whole question and see if you arrive at the answer *you* expect. (This should not take long at all using an electronic calculator. You already have the steps for the calculation; just enter the new set of values.)

9. If you are acquiring these skills for examination purposes, do some regular work over as many weeks as possible. That way the system will *grow with you*. Do not cram stuff new stuff, particularly new methods, in the last few hours leading to your exams.

10. Wherever a reference to another question is provided, please look up that question as soon as possible (and, if necessary, work through it). Working on similar or contrasting material is a very effective way of reinforcing your understanding and authority.

11. *Do not memorise the steps shown in the solutions. That would defeat the whole purpose of this book.* Remember, you should avoid dependence on memorised material. Instead, you should develop your deductive skills. This can only be achieved by feeling, understanding, and deducing, not by memorising.

[The intention of the author is that you (the student) become an organised, clear-thinking, authoritative attacker of biochemical calculations (or anything quantitative); that you no longer need to try to recall what your lecturer said; that you no longer need to try to recall formulae; that you do not need to try to recall what's in this book; that you, by yourself, are able to figure things out; that you believe and know that you are on the right track towards solving the calculation (and you are certain when you have arrived at the correct answer) — now (when you are learning from this book), at your exams, next year, and forty years from now!]

2
Units and Amounts

SI units (*Système International d'Unités*) are the preferred units used in chemistry and biochemistry. The relevant ones in biochemistry are shown in Table 2.1.

In biochemistry, it is more acceptable to be a bit more clear and descriptive than to be too strict about definitions and symbols. It is perfectly all right to write *mole* or *sec* (instead of the SI symbols *mol* or *s*) and even use calories (cal) in calculations (for amount of heat energy). We do have some units that are used in biochemistry in conjunction with those in Table 2.1 and are likely to be retained because of their convenience. There are a number of units that are used in biochemistry that relate to water (like °C and calories). However, the current trend is to go all the way using SI units. (A full set of units and their conversions that may be of interest to biochemistry students is given in Appendix 1.)

When submitting research papers for publications, authors are strictly required to follow the "Instructions to Authors" provided by the editors of the journals. Unfortunately, each journal has different requirements, and there are local (or country) preferences. For example, M is tending to be reserved for the prefix "mega" (see Table 2.2) and not for "molar" (as it was used until quite recently); however, most biochemistry journals still permit its use for "molar." Both upper case L and lower case l are also permitted for litres by most journals. Gallons, pints, and fluid ounces are still used in many advanced countries, although all use SI units in international science journals. Students should endeavour *to be able to* convert quantitative information into units and measures, and even utensils that ordinary folks use—for example, knowing how to dilute a herbicide for farm use, knowing the volume of a household bucket, and being aware that the volume contained in a teaspoon is 5 mL (hence liquid pharmaceuticals concentrations are often expressed in amount per 5 mL).

Use of Prefixes

Units may be very large or small, and writing them can involve many zeros (e.g., 432 000 000 or 0.000 000 432); if the zeros are not conveniently grouped, it can be difficult to gauge the magnitude of the number. It is true we can write them in **scientific notation** (e.g., 4.32×10^8 or 4.32×10^{-7}). If one is not constantly dealing with scientific notations, there might be a lack of feel for numbers expressed this way. In biochemistry, instead of using the whole range of powers or exponents (of base 10), only 10^6, 10^3, 10^{-3}, 10^{-6}, 10^{-9}, 10^{-12}, 10^{-15}, etc., are used, and these multiples, or fractions or "submultiples," are called by prefix names; they are SI approved and preferred by most journals.

Table 2.1	SI Units Relevant to Biochemistry	
Physical Quantity	**Name**	**Symbol**
Length	Metre	m
Mass	Kilogram	kg
Time	Second	s
Chemical amount	Mole	mol
Energy	Joule	J
Thermodynamic temperature	Kelvin	K
Volume	Cubic metre	m^3

Table 2.2	Prefixes for Multiples and Fractions of Units	
Factor	**Prefixes**	**Symbol**
10^6	mega	M
10^3	kilo	k
10^{-3}	milli	m
10^{-6}	micro	μ
10^{-9}	nano	n
10^{-12}	pico	p
10^{-15}	femto	f
10^{-18}	atto	a

In biochemical calculations, some awkward situations may arise where a scientific notation (number) occurs in combination with a prefixed unit (e.g., 3.42×10^{-5} μmol). (For advice on how to best express these numbers and a general review of exponentials, please see Box 2.1.) When expressed with prefixes, the values provide a much better feel: for example, "Glucose occurs in millimolar quantities in blood" (4.4 mmol/L), or, "μg amounts of DNA can be analysed." This follows from the fact that biochemists tend to use a **sensible range of numbers.** That range is 1 to 999 (see Box 2.1 for further comments). For this range, do not use scientific notation. For example, write 232 (not 2.32×10^2) and write 0.14 (not 1.4×10^{-1}). Scientific notation should not be seen as an impressive way of expressing numbers.

For consistency in tables, the following expressions are acceptable:

$$0.03 \times 10^{-2}$$

$$2.32 \times 10^{-2}$$

$$9.76 \times 10^{-2}$$

$$47.63 \times 10^{-2}$$

Box 2.1 Review of Exponentials

In the broad meaning, the word "exponential" refers to anything "involving an exponent or power term," for example, 2^3, 10^6, 2.4×10^7, 24×10^7, 6.022×10^{23}, $10^{0.3010}$, x^y, or e^x. As with most mathematical expressions, if not otherwise stated, 1 is understood. So 2^3, 10^6, $10^{0.3010}$, x^y, and e^x become, respectively, 1×2^3, 1×10^6, $1 \times 10^{0.3010}$, $1 \times x^y$, and $1 \times e^x$. Having written the exponentials in full exponential notation form, we can now identify the different parts of these expressions:

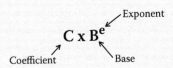

The exponent may also be negative. This tells us that the entire B^e part of the expression is to be considered as a reciprocal:

$$2.4 \times 10^{-7} = 2.4 \times \frac{1}{10^7} = \frac{2.4}{10\,000\,000} = 0.000\,000\,24.$$

(This is the reason why dividing by 10^{-7} is the same as multiplying by 10^7.)

In more exacting meanings, the word "exponential" is used in the terms "exponential function" or "exponential notation." (Exponential functions are e^x, where e is the base of natural logarithms and x is a variable. Exponential functions are not our interest here.) Exponential notation is writing numbers in the form $C \times B^e$. (The ordinary way of writing numbers is called the "standard notation" — for example, 47, 478 000, 478.47, 0.0056, etc.) In exponential notations, the coefficient, the base, and the exponent may take *any* value (yes, even the base can be negative!) All numbers can be written in all sorts of exponential forms. There are two special situations: when the base is 10, we call the exponent the *logarithm* (or, log to the base ten), and when the base is **e** (= 2.718), we call the exponent the *natural logarithm* (ln).

The log of 1 is 0 and the log of 10 is 1 (100, 2; 1000, 3; etc.). Decimal values were worked for numbers between 1 and 10 and meticulously listed in the so-called "four figure log tables." To fully appreciate logs, they must be written as exponents: for example, $200 = 10^{2.3010}$. It was in this context, we had the words *characteristic* referring to the whole number, 2, and *mantissa* to the decimal part. Logs were extensively used as an aid to arithmetic operations in the preelectronic calculator days. While electronic calculators have replaced that role, logarithms, both natural and the base-ten variety, remain of interest to mathematicians. We, as biological scientists, need logarithms to fully understand concepts such as exponential growth, radioactive isotope decay and half-lives.

Scientific notation is a convenient way of writing very large or very small (standard notation) numbers. For example:

$$432\,000\,000 = \mathbf{4.32 \times 10^8}$$

$$0.000\,000\,432 = \mathbf{4.32 \times 10^{-7}}$$

With scientific notation, the coefficient must be a single-digit (positive or negative) integer, the base positive 10 and, as discussed before, the exponent may be a positive or negative integer.

Levels of precision, as "number of significant figures," may be included in the coefficient, as in 4.32×10^8 or 4.3276×10^8. If the coefficient is not a single-digit integer, then these expressions are simply called "exponential notations," (e.g., 345×10^5, 0.23×10^3).

With scientific notations, the exponent may be any integer. In the SI-preferred system of writing numbers, only the "third" integer is permitted (i.e., 10^3, 10^6, 10^9, etc. and 10^{-3}, 10^{-6}, 10^{-9}, etc.). (And these levels or magnitudes of numbers, when used with units, are referred to by prefixes, such as milli-, micro-, nano-, etc.). With this restriction placed on the exponent, the coefficient must be allowed a wider range. The range allowed is 1 to 999. However, it is common practice in biochemistry and chemistry to include simple decimal numbers as acceptable "coefficients" for prefixed units. For example, it is acceptable to write 0.2 μmol, 0.5 μmol, or 0.23 μmol, but 0.003 μmol is often written as 3 nmol and 0.987 μmol as 987 nmol. We have always written 0.5 M HCl (not 500 mM HCl). The rule of thumb seems to be: *if the decimal is a "simple" one, use it.*

One other problem arises in biochemical calculations. In certain calculations, it is possible for an expression containing a scientific notation and a prefixed unit to appear together (e.g., 3.42×10^{-5} μmol). Clearly, this is an awkward expression and needs to be "converted." (The value of this example is best shown as pmol. For more examples dealing with this kind of problem, please see Q2.17.)

The effect created here is to show the increase in magnitude, so all numbers are kept at 10^{-2}. Suppose the above figures represent concentrations in molar (mol/L); it would be best to express them as mmol/L and, if part of a table, present them as shown below.

mol/L
0.3
2.3
97.6
476.3

If you are hand writing symbols, exercise care: **M** is different from **m** and **μ** has a long tail, etc. If you are using a word processor, spend time to select and use the absolutely correct symbol.

Mass

Gram (g) is the elementary unit of mass mainly because the kilogram (actually the base unit in SI, in spite of it containing a prefix!) is too large. Students are encouraged to use mg (10^{-3} g), μg (10^{-6} g), and so forth.

Mole (mol) is a unit expressing an *amount* of chemical substance. It is relative molecular mass (M_r) or molecular weight (MW) expressed in grams.

It is the author's firm belief that this definition is the most meaningful and certainly the most practical. Students who have been taught that mole is a number, or other versions, may consider all of them as *properties* of a mole: One mole of a substance contains Avogadro's number (6.02×10^{23}) of molecules. Or, one mole of a substance contains the same number of molecules as there are atoms in 12 g of carbon.

$$1 \text{ mole glucose } (M_r \text{ } 180) = 180 \text{ g}$$

$$1 \text{ mole glucose dehydrogenase } (M_r \text{ } 1\,000\,000) = 1\,000\,000 \text{ g} = 1000 \text{ kg}$$

Relative molecular mass (M_r), or **molecular weight (MW)**, was also known as **formula weight (FW)**. It is important to realise that M_r (MW or FW) is a number. It is the number of times one molecule of a substance is heavier than 1/12th of a carbon atom (or approximately heavier than one atom of hydrogen). Note, M_r, MW, and FW have no units, although **amu (atomic mass units)** was once created for those who like to put something next to numbers. Note, the currently preferred term and symbol is relative molecular mass (M_r). This should reinforce the notion that it is number, indicating the molecule's mass relative to 1/12th of a carbon atom. It is therefore dimensionless.

[Also please note, with the change of preference to relative molecular mass (M_r) over molecular weight (MW), **relative atomic mass (A_r)** now has preference over **atomic weight (AW)**.]

Dalton (Da), on the other hand, is defined as a unit of mass. It is similar to the once-used amu. One dalton equals 1.661×10^{-24} g. A single atom of ^{12}C is defined to have a mass of 12 daltons, therefore

$$1 \text{ dalton} = \frac{1 \text{g}}{\text{Avogadro's No.}}$$

It is best to restrict the use of daltons to only certain biochemical references. Do not use daltons with molecules whose make-up is simple or uncomplicated. Instead, use daltons to indicate relative sizes of biological complexes or aggregates. The term is most useful for referring to structures such as chromosomes, ribosomes, mitochondria, electron transport complexes, and whole cells where the term "molecule," and therefore M_r, would be inappropriate.

For example, the mass of one cell of *Escherichia coli* is about:

$$1 \times 10^{-12} \text{ g or 1 pg or } \frac{1 \times 10^{-12}}{1.661 \times 10^{-24}} = 0.6 \times 10^{12} \text{ daltons}$$

A dalton (Da) is really a very small fraction of a gram. It is about the mass of an atom of hydrogen. (Daltons are extensively used in molecular biology [see Chapter 13]). In this context, lately, the term **molecular mass** is gaining more and more usage. (In disciplines other than biological sciences, the once discarded amu has been reinvented and formalised as the ***unified* atomic mass unit**, given more trendy symbol, u, and gained SI approval for use with SI units.) Daltons and unified atomic mass units are exactly the same thing; they are both units of mass.

Note, a protein may be said to have a relative molecular mass of 50 000 (M_r = 50 000) or a molecular mass of 50 000 Da (more conveniently, 50 kDa) and may be referred to as the 50 000-M_r protein or the 50 kDa protein. It is not correct to express M_r in Daltons. It is also not acceptable to use K or k to represent 1000 in M_r. (Write M_r = 50 000; do not write M_r = 50K nor M_r = 50k.)

Volume

For liquid measurements in biochemistry, use **litres** and not 1×10^{-3} m^3 (its equivalent) or m^3 (the base unit in SI) or dm^3 (which equals 1 litre, and the original definition of a litre). Use mL and not cm^3 or cc. Litres are usually used as a measure of liquids (fluids generally), whereas the cubed dimensions, such as cm^3 and m^3, are used as measures of space. It is certainly litres in biochemistry. While mL is the most commonly unit in undergraduate biochemistry laboratories, μL is also measurable quantity with the use of micropipettes and injection systems that are part of modern instruments.

[**Comment on Spelling and Symbols:** The spelling is "liter" in the U.S. and "litre" in British English-speaking countries, including Australia. Both the upper case "L" and the lower case "l" are permitted by British and U.S. science journals. Some countries, including the U.S. and Australia, officially use the upper case L in all commercial and everyday usage. The recommendation is: wherever permitted, use the uppercase "L" because the lowercase "l" is known to cause confusion with the number 1 and the capital letter I.]

Interconversions of cubic measure can be tricky. Consider the following:

$$1000 \text{ L} = 1 \text{ m}^3$$

$$1 \text{ L} = 1 \text{ dm}^3 = 10^{-3} \text{ m}^3$$

$$1 \text{ mL} = 1 \text{ cm}^3 = 10^{-6} \text{ m3}$$

$$1 \text{ } \mu\text{L} = 1 \text{ mm}^3 = 10^{-9} \text{ m3}$$

Students should attempt to visualise the size of a millilitre (mL) and microlitre (μL) by realising that:

$$1 \text{ mL} = 1 \text{ cm}^3 \text{ (a cube of 1 cm)}$$

$$1 \text{ } \mu\text{L} = 1 \text{ mm}^3 \text{ (a cube of 1 mm)}$$

Note: dm is a *decimetre*, or one-tenth of a metre, or 10 cm. Besides realising that 1 L equals 1 dm³, the usage of decimetres (dm) is discouraged (as *deci-* is not one of the SI preferred prefixes).

Concentrations

Broadly speaking, two kinds of concentration expressions are used in biochemistry. Gravimetric, which is based on mass or weight, and chemical or molar concentration, in which the amount of solute is expressed in moles.

Gravimetric Concentration

Commonly used gravimetric concentration units in general and clinical biochemistry are:

$$g/100 \text{ mL} = g/dL$$

$$mg/mL$$

$$g/L$$

$$mg/100 \text{ mL} = mg/dL$$

Except for chemists expressing density as g/mL, this term is not commonly used by biochemists.

While considerable uniformity — based on the SI system — is being introduced internationally by authorities in the medical and health sciences, unfortunately variations still exist. Students who are enrolled in professional disciplines (of studies) are advised to become familiar with and use their professional bodies' approved or recommended set of units as early as possible in their programs.

[Decilitre, dL, is a relatively popular reference volume in health sciences in many countries. A decilitre is one-tenth of a litre. There is potential for confusion, with litre being defined as a cube of a decimetre, itself one-tenth of a metre. Science authorities encourage the use the SI-preferred prefixes such as *milli-* and *micro-* (and the avoidance of *deci-* and *centi-*). The author, therefore, believes that the term "decilitre" should be phased out. If a litre is considered too large as a reference volume and a mL is too small, there is a convenient replacement (and equivalent of decilitre), 100 mL.]

Percentage (%) is an expression of gravimetric concentration. Although it causes considerable confusion, it is still around and biochemistry students should be able to handle concentrations expressed in this manner.

A 5% solution of acetic acid in water could mean:

- 5 g acetic acid per 100 g water
- 5 g acetic acid per 100 mL water
- 5 mL acetic acid per 100 mL water.

As an attempt to overcome the ambiguity, units such as % (w/w), % (w/v), and % (v/v) (respectively, weight to weight, weight to volume, and volume to volume) are used. Even then, the question arises: is it 5 g of substance in a final weight of 100 g of solution or 5 g of substance in 100 g of water (the wording really means 5 g in [already measured out] 100 g of water — i.e., 5 g of substance *plus* 100 g water)? The latter might give a final volume greater than 100 mL.

Students must realise that practically all solution preparation in biochemistry involves *making up to the final volume* with the solvent and *not* adding a fixed volume of solvent.

The dictionary meaning of "percent" is parts per 100 parts. This meaning is not evident when 5 g of a substance is contained in a final weight of 100 g solution. % (w/w) is taken to mean g/100 g final mass of solution. % (w/v) is taken to mean g/100 mL final volume of solution. For *dilute* solutions, it does not matter whether it is 100 g or 100 mL because 100 mL pure water weighs very close to 100 g; such would be the case with 0.2% (w/w) glucose. However, it does matter with concentrated solutions. Concentrated sulfuric acid (H_2SO_4), for example, is 95% (w/w) H_2SO_4 (there are 95 g H_2SO_4 in 100 g final mass of solution). Now, this solution of H_2SO_4 *is* heavy: 100 mL weighs 183 g. In the exercises that follow, students will note that density will have to be considered in such situations.

Of biochemical interest, students may wish to note that glucose concentration in blood is sometimes mentioned as 80 mg %. The meaning here is 80 mg glucose per 100 mL blood. The blood weighs a lot more than 100 g. (*Note:* "mg %" is not an accepted unit.)

Parts per million (ppm) is another gravimetric concentration or proportion unit. Its use should be restricted to indicating minute amounts — up to around 200 ppm. It is a term used by environmental scientists, toxicologists, pharmacists, and geologists. Alternative terms to ppm are:

$$ppm = \mu g/mL = mg/L$$

$$ppm = \mu g/g = g/tonne \text{ (one tonne} = 1000 \text{ kg).}$$

These alternative terms provide more clarity. $\mu g/mL$ is preferred when a very small amount of matter is present in a liquid (usually water). A pharmacist would mean $\mu g/g$ when mixing a medication into a paste, whereas a geologist would mean g/tonne when reporting the amount of metal in a tonne of rock. A toxic gas expressed in ppm would have to be volume of the toxic gas per million volumes of air. There are some interesting gas laws relating to molar gas volumes and proportions of gases in mixtures that students are advised to review. Students may also wish to verify that 1 ppm = 0.0001%.

Percent saturation. Protein purification often involves precipitation with neutral salts. These methods precipitate protein in the native (not denatured) state. Ammonium sulphate is most commonly used and in very high, or saturating, concentrations. Globulins and albumins owe their definition to this precipitation method of separation.

The term "100% saturation" means the maximum possible concentration of the salt at the given temperature. Based on this unit for expressing concentrations, if one volume of saturated ammonium sulphate (SAS) is mixed with one volume of water (usually the water containing protein), the new concentration is 50% saturation of ammonium sulphate.

Preparation of SAS usually involves dissolving as much ammonium sulphate as possible at a temperature a few degrees higher than room temperature, allowing the solution to cool to room temperature, and using the clear liquid. (Why should SAS not be stored in the fridge?)

Molarity

One molar (mol/L) solution is a solution containing one mole of substance in one litre (final volume) of solvent.

$$1 \text{ molar} = 1 \text{ mol/L} = 1 \text{ mmol/mL}$$

$$1 \text{ millimolar} = 1 \text{ mmol/L} = 1 \text{ } \mu\text{mol/mL}$$

$$1 \text{ micromolar} = 1 \text{ } \mu\text{mol/L} = 1 \text{ nmol/mL}$$

Note: mol/L is the new replacement unit for M.

Students should check that they gain a clear understanding of the above relationships. (This understanding makes it much easier to visualise steps in biochemical calculations.) Most important is μmol/mL because biochemistry commonly deals with mL volumes of mmol/L concentrations. Also, note carefully that mole is an amount and molarity is a concentration. Molarity includes an amount unit (moles) and a volume unit (litres). This is reinforced in the current trend to use mol/L instead of M.

Normality (N) is an older concentration term used in expressing concentrations of acids, alkalis, and compounds involved in oxidation-reduction. Normality and related terms apply to half-equations. One **normal** solution contains one **equivalent** of solute per litre. An **equivalent weight (EW)** may be the same as the relative molecular mass, or a simple fraction of the relative molecular mass.

$$M_r \text{ H}_2\text{SO}_4 = 98 \quad \text{EW H}_2\text{SO}_4 = \frac{98}{2} = 49$$

$$M_r \text{ HCl} = \text{EW HCl} = 36.5$$

$$M_r \text{ NaOH} = \text{EW NaOH} = 40$$

The equivalent weight is the formula weight that contains one mole replaceable hydrogen or hydroxyl ion (or mole-electrons in the case of oxidation-reduction). **Equivalent** is a term corresponding to mole; it is equivalent weight in grams.

$$\text{One equivalent H}_2\text{SO}_4 = 49 \text{ g}$$

$$\text{One equivalent HCl} = 36.5 \text{ g}$$

Normality (the concentration term) relates to molarity (mol/L) as follows:

$$1 \text{ mol/L } H_2SO_4 = 2 \text{ N } H_2SO_4 = 98 \text{ g/L}$$

$$1 \text{ mol/L HCl} = 1 \text{ N HCl} = 35.5 \text{ g/L}$$

Check this: 1 mol/L H_2SO_4 is twice as concentrated as 1 mol/L HCl with respect to acidity. Remember monobasic and dibasic acids? Normality and equivalents can be useful in calculations involving dibasic acids.

Density and Specific Gravity

The common unit of **density** of solutions in chemistry is grams per millilitre (g/mL). While density is the mass of one unit volume, **specific gravity** is simply the density of a substance *compared* with the density of water. It follows that specific gravities have no units; they are just numbers. In physics, density would be grams per cubic centimetre (g/cm³ or g cm⁻³); in SI, it is kilograms per cubic metre (kg m⁻³).

Biochemistry students must exercise caution in the use of expression of units in power terms (exponential notations) — in other words, terms such as g cm⁻³. The author believes that, because "per" or "/" (such as g/cm³ or cm/sec) provides better feel for most students, these expressions should be used (particularly by non-physics science students). Do not use power terms purely to impress. There are some inherent difficulties in the use of exponent/power-containing expressions. Consider terms such as m sec⁻¹ sec⁻¹ or m⁺¹ sec⁻¹. The former is an acceleration unit, and the author believes that it would be misleading if written as m sec⁻². Compare this with the units for force per unit area, which is Newtons per square centimetre, N cm⁻². To write m⁺¹ instead of simply m is totally unnecessary. (Experienced physicists may use it *while performing* exponential-factor-label method of calculations.)

Numbers

The rules that apply to the writing of numbers are similar to those that apply to the writing of symbols. In addition to conventions being followed, proper practices avoid ambiguities and mistakes by the writer or reader and they create better feel of the magnitude of the numbers. Guidelines for writing numbers are listed here:

1. Write all decimal numbers less than one by starting with a zero: write 0.5, never .5. In other words, do not write numbers with a "naked" decimal point. Develop this habit when recording readings in the laboratory, in your calculations, and in final answers. (Electronic calculators and spreadsheet programs have made us lazy in this regard because they automatically enter the zero for us. We have to be careful in other situations when it is not automatically done.)

   The number .5 can and has been interpreted as 5, leading to medication-dosage errors, for instance. The pharmaceutical and medical authorities have very strong objections to this, as they do to even writing numbers with a "trailing zero." An example of a number with a trailing zero is 1.0 (which could be mistaken for 10). As science students, we use trailing zeros to convey precision with "number of significant figures." However, the authorities argue that the occasional disastrous consequence outweighs the desire to show precision. Once students are aware of this problem, appropriate measures can easily be taken to avoid ambiguities. (Consider perhaps writing *1 mg exactly*, instead of *1.0 mg*, where necessary. Students may continue to use numbers expressed as 2.0 or 2.00, especially in intermediary steps in their calculations, or if number of significant figures is indeed what they wish to demonstrate.)

2. Commas are no longer used but digits are grouped in triplets (counting either direction from the decimal point).

This can however cause problems in word processing programs. If you are typing the above number towards the end of a line or in narrow columns, the number could break up as:

47 328.610

24

or

47

328.610

Watch for it and avoid it.

3. Be aware that certain European and South American countries use commas instead of decimal points.

Multiplication and Division Errors

Even at second-year university level, some students fall prey to simple calculation errors. Please study the statements below:

- Dividing by a number > 1 creates a smaller number.
- Dividing by a number < 1 creates a bigger number.
- Multiplying by a number > 1 creates a bigger number.
- Multiplying by a number < 1 creates a smaller number.
- Multiplying by 10^6, creates a bigger number.
- Multiplying by 10^{-6} creates a smaller number. This is actually dividing by 10^6.
- While divisors may be written as multiplicand (raised to a negative exponent), avoid this with simple numbers because it deprives you of the feel that the answer is getting smaller. Write 56/1000, not 56×10^{-3}.
- Students are advised to keep their thinking consistent with what they learned in primary school: If you divide, the number gets smaller; if you multiply, the number gets bigger.

(Of course, you did not know then about negative numbers, decimals, and exponents — certainly not negative ones.) Keep the consistency going! That will give you better feel (and more confidence in evaluating answers or steps in calculations). If you do multiply by 10^6, the number will get bigger — for example, 5 μmol $\times$ 10^6 = 5 000 000 μmol. The *number* has become bigger, but then you might decide to change the units and call it 5 mol — that's a different matter. Likewise if you divide by 10^6, the number will get smaller: 7g $\div$ 10^6, or (written) $\frac{7}{10^6}$, becomes 0.000 007 g). If for operational reasons or convenience, you chose to write 7×10^{-6} instead of writing $\frac{7}{10^6}$, that, too, is a different matter. Your thinking should remain consistent: that the number does get smaller. Indeed your thinking should be there even if you hardly ever use the $\div$ sign; writing your number below the horizontal fraction bar, or *vinculum*, means division.

Developing Confidence with Units

All biochemical calculations involve units. Throughout this book, students are shown methods that claim better understanding, feel, confidence and, eventually, authority over biochemical calculations. Units are an integral part of the calculations. It follows, therefore, that to gain confidence over units is to gain a considerable advantage over the calculations.

There are, as discussed earlier in this chapter, two sets of units that are commonly used in biochemical calculations:

1. Amount or mass units (gravitational or chemical)
2. Concentration units (gravitational or chemical).

Students are advised for each of the four kinds of units to pick one unit and to relate all other derivatives (multiples or fractions) to that base or fundamental unit. For gravitational amount (mass), use g as the base unit. (For this exercise, please do not worry about kg being the "base unit" in SI.) If you are faced with a number in "mg," immediately think of it as "one-thousandth of a gram." Likewise, think of "μg" as one-millionth of a gram, and so forth. By adopting this approach, 3876 mg can easily be converted to 3.876 g, and vice versa. Do these conversions "by inspection." Remember that in your junior school days, you learned to do this by shifting the decimal point. Now, you can do this ever so confidently. Do not even think of dividing or multiplying by 1000, or even worse, 10^3!

Use mol as the base unit for chemical amounts, and proceed in the same way as above.

A different approach should be taken with concentrations. With gravimetric concentrations, again stick to g as the base mass unit, but you may think of L or 100 mL or 1 mL for your volume. It is not difficult, for your mind's eye to see any one of the following expressions, in either of the two other forms just by staring at it:

$$32 \text{ g/L} = 3.2 \text{ g/100mL} = 32 \text{ mg/mL}$$

(Again it is the shifting-the-decimal-point trick.)

For chemical concentrations, use mol/L as your base unit and learn to quickly recognise:

$$1 \text{ mol/L} = 1 \text{ mmol/mL}$$

$$1 \text{ mmol/L} = 1 \text{ μmol/mL}$$

$$1 \text{ μmol/L} = 1 \text{ nmol/mL}$$

With this kind of approach, you should be able to visualise that 1 mL contains one-thousandth the amount contained in 1 L.

If you adopt the suggestions above, they will become second nature to you and you will be able to perform certain conversions intuitively. That could mean doing less steps in your calculations. With fewer steps, there is less chance of making mistakes. (In some exams, you may be required to "show your working." This does not mean that you have to show working for writing "34 μmol/L = 34 nmol/mL"!)

Questions

1. A solution of valine (M_r 117) contains 2.94 g/L.
   (a) Calculate its molarity in mmol/L.
   (b) How many μmol are there in 5 mL of this solution?

2. A solution of arabinose (M_r 150) contains 38.7 mg/100 mL.
   (a) Calculate its molarity in mmol/L.
   (b) How many μmol are there in 1.0 mL of this solution?

3. A solution of a protein (M_r 27 000) was made containing 2.8 g per 400 mL.
   (a) What is the μmol/L concentration of this solution?
   (b) How many μmol are there in 10 mL of this solution?

4. Consider a 2.27 mmol/L solution of a substance (M_r 590). Fill in the blanks.
   (a) μmol/mL   = _____
   (b) g/L   = _____
   (c) % (w/v)   = _____
   (d) volume (in mL) containing 20 μmol   = _____

5. Consider a 1.8 mmol/L solution of a protein (M_r 27 000). Fill in the blanks.
   (a) μmol/mL   = _____
   (b) g/L   = _____
   (c) mg/mL   = _____
   (d) volume (in mL) containing 10 μmol   = _____

6. Glucose concentration in blood has been referred to as both 4.4 mmol/L and as 80 mg %. Are these values equal? M_r glucose 180

7. A 0.2 g/100 mL solution of a purified protein is determined to be 72 μmol/L.
   (a) How many μmol are there in 15 mL of this solution?
   (b) Calculate M_r of the protein.

8. An aqueous solution of 95.0% (w/w) H_2SO_4 has a density of 1.83 g/mL. Calculate the molarity of the solution.
   A_r H 1 O 16 S 32

9. How many moles of sodium chloride are there in 1 L of a 0.9% (w/w) solution having a density of 1.1 g/mL?
   A_r Na 23 Cl 35.5

10. Assume that a 40% (w/w) solution of a protein (M_r 9 000) in water has a density of 1.20 g/mL. What volume of this solution would you need to take in order to have 100 mmol?

11. Assume that a 30% (w/w) solution of a protein in water has a specific gravity of 1.20. 2.5 mL of this solution is analysed to contain 120 μmol. Calculate the M_r.

12. An enzyme (M_r 30 000) is available commercially as 0.2 g/100 mL at 80% purity. What volume (in whole mL) would you take to have around 0.1 μmol enzyme needed for an assay?

13. Assume that an invertebrate tissue (90% water) contains 500 different enzymes of average M_r 100 000. The enzymes represent 0.7% of the wet weight of this tissue. Also, assume that each enzyme is of equal concentration. Calculate the μmol/L concentration of each enzyme.

14. A 50% (by volume) solution of ethanol is prepared by mixing equal volumes of ethanol with water; this solution has a density of 0.93 g/mL. (M_r ethanol 46, density 0.79 g/mL) Calculate:
    (a) % (by weight) of the ethanol solution
    (b) molarity (mol/L) of the ethanol solution

15. 40 ng of a 50 kDa protein is dissolved in 100 mL. Calculate the pmol/L concentration of this solution.

16. Consider a 0.8 μmol/L solution of a 60 kDa protein. Calculate the μg mass of protein in 1.0 mL.

17. Convert the expressions in Table 2.3 into SI-preferred prefixed units (correct to 3 significant figures). For example, 3.673×10^{-5} mol becomes 36.7 μmol.

18. Consider a 90 kDa protein. Perform the necessary conversions and write your final answers in Table 2.4.

Your final answers must meet the following requirements:

- must be within the range 1 to 999
- must have an accuracy of three significant figures
- must show the correct prefix (for the unit you have converted into)

For example, if the answer is 3.673×10^{-5} mol, write the final answer as "36.7 μmol" in Table 2.4.

Table 2.3   Blank Table for Question 2.17

(a) 0.0397 mmol/L	
(b) 392.78×10^{-2} g	
(c) 4.48×10^{-3} μmol	
(d) 3.677×10^{-5} μmol	
(e) 1.904×10^{-5} mol	
(f) 27.04×10^{-4} nmol	
(g) 380.4×10^{-1} nnol	
(h) 4304×10^{-2} pmol	

Table 2.4   Blank Table for Question 2.18

	Enter (3 sig. figures)	Enter Prefix	unit
(a) 80 μg			mol
(b) 460 pg			mol
(c) 80 mmol			g
(d) 460 nmol			g
(e) 80 μg/L			mol/L
(f) 460 mg/L			mol/L
(g) 20 mL 80 μmol/L			g
(h) 20 mL 460 μmol/L			g

Answers

1. (a) Molarity $= \dfrac{2.94}{117} = 0.0251$ mol/L $= \mathbf{25.1\ mmol/L}$

   (b) 25.1 mmol/L = 25.1 μmol/mL

$$5 \text{ mL contains } 25.1 \times 5 = \mathbf{125.5\ \mu mol}$$

2. (a) 387 mg in 1 L

$$\text{Molarity} = \frac{387}{150} = \mathbf{2.58\ mmol/L}$$

   Note: Mass is in mg, concentration comes out directly in mmol/L.

   (b) **2.58 μmol** in 1.0 mL

   Note: You should be able to do part (b) "by inspection."

3. (a) 0.4 L contains $\dfrac{2.8}{27000}$ mol

$$1 \text{ L contains } \frac{1}{0.4} \times \frac{2.8}{27000} \text{ mol} = 0.000\ 259 \text{ mol/L}$$
$$= 259 \text{ μmol/L}$$

   (b) 1 mL contains 0.259 μmol

$$10 \text{ mL contains } \mathbf{2.59\ \mu mol}$$

   Note: Again possible by inspection: dividing part (a) answer by 100. Or, this thinking can be used for evaluation: answer to (b) should be 1/100th of (a).

4. (a) 2.27 mmol/L = **2.27 μmol/mL**
   (b) 2.27 mmol/L = 2.27 × 590 mg/L

$$= 1339.3 \text{ mg/L}$$
$$= \mathbf{1.34\ g/L}$$

(c) 1.3393 g/L = **0.134 g / 100 mL = % (w/v)**

(d) 2.27 μmol in 1 mL

$$20 \text{ μmol in } \frac{20}{2.27} \times 1 = \textbf{8.81 mL}$$

5. (a) 1.8 mmol/L = **1.8 μmol/mL**

(b) 1.8 mmol/L = 1.8 × 27 000 mg/L

$$= \textbf{48.6 g/L}$$

(c) 48.6 g/L = **48.6 mg/mL**

(d) 1.8 μmol in 1 mL

$$\Rightarrow 10 \text{ μmol in } \frac{10}{1.8} \times 1 \text{ mL} = \textbf{5.56 mL}$$

6. "80 mg %" is supposed to mean 80 mg /100 mL = 800 mg/L

$$4.4 \text{ mmol/L} = 4.4 \times 180 = 792 \text{ mg/L}$$

Answer: Yes, close enough.

7. (a) 1000 mL contains 72 μmol

$$\Rightarrow 15 \text{ mL contains } \frac{15}{1000} \times 72 \text{ μmol}$$

$$= \textbf{1.08 μmol}$$

(b) 0.2 g/100 mL = 2 g/L

$$72 \text{ μmol} = 2 \text{ g}$$

$$72 \text{ mol} = 2 \times 10^6 \text{ g}$$

$$1 \text{ mol} = \frac{1}{72} \times 2 \times 10^6 \text{ g}$$

$$= 27\,778$$

$$\text{i.e., } \textbf{M}_r = \textbf{28 000}$$

8. 100 g solution contains 95 g H_2SO_4

$$100 \text{ g solution contains } \frac{95}{98} \text{ mol } H_2SO_4$$

$$100 \text{ g solution occupies a volume of } \frac{100}{1.83} \text{ mL}$$

$$\text{i.e., } \frac{100}{1.83} \text{ mL contains } \frac{95}{98} \text{ mol } H_2SO_4$$

$$\Rightarrow 1000 \text{ mL contains } 1000 \times \frac{1.83}{100} \times \frac{95}{98}$$

$$= 17.7 \text{ mol}$$

$$\textbf{Molarity} = \textbf{17.7 mol/L}$$

9. 100 g solution contains 0.9 g NaCl

$$100 \text{ g solution contains } \frac{0.9}{58.5} \text{ mol NaCl}$$

$$100 \text{ g solution occupies a volume of } \frac{100}{1.1} \text{ mL}$$

$$\text{i.e., } \frac{100}{1.1} \text{ mL contains } \frac{0.9}{58.5} \text{ mol NaCl}$$

$$\Rightarrow 1000 \text{ mL contains } 1000 \times \frac{1.1}{100} \times \frac{0.9}{58.5}$$

$$= \mathbf{0.169 \ mol}$$

10. $1 \text{ mL contains } \frac{40}{100} \times 1.20 \times \frac{1}{9000} \times 10^6 \text{ } \mu\text{mol}$

$$= 53.3 \text{ } \mu\text{mol}$$

$$\Rightarrow 100 \text{ } \mu\text{mol in } \frac{100}{53.3} = \mathbf{1.88 \ mL}$$

11. 1 mL weighs 1.2 g

$$2.5 \text{ mL weighs } 2.5 \times 1.2 \text{ g}$$

$$2.5 \text{ mL contains } 2.5 \times 1.2 \times 0.3 \text{ g protein}$$

$$120 \text{ } \mu\text{mol} = 2.5 \times 1.2 \times 0.3 \text{ g}$$

$$1 \text{ } \mu\text{mol} = \frac{1}{120} \times 2.5 \times 1.2 \times 0.3 \times 10^6 \text{ } \mu\text{g (see note below)}$$

$$= \mathbf{7\,500 \ (= M_r)}$$

Note: $\text{mol} = M_r \text{ in g}$

$\text{mmol} = M_r \text{ in mg}$

$\mu\text{mol} = M_r \text{ in } \mu\text{g}$

12. $\dfrac{0.2}{3 \times 10^4} \text{ mol in 100 mL}$

$$\Rightarrow 0.1 \times 10^{-6} \text{ mol in } 0.1 \times 10^{-6} \times \frac{3 \times 10^4}{0.2} \times 10^2 \text{ (see comment 1 below)}$$

$$= 1.5 \text{ mL}$$

This, if 100% pure.
Since only 80% pure, we need more than 1.5 mL (see comment 2 below)

$$= 1.5 \times \frac{100}{80} \text{ mL}$$

$$= 1.875 \text{ mL}$$

$$= \mathbf{2 \ mL} \text{ (in whole mL)}$$

Comments

1. Take a good look at this expression (that is keeping track of what is going on or evaluating). You may wish to do only:

$$\frac{0.1 \times 3}{0.2} = 1.5$$

You may avoid entering $10^{-6} \times 10^4 \times 10^2$ in your calculations because they cancel out. Entering them, particularly as exponentials, may give rise to errors. (A common mistake is the failure in creating an operational negative exponential.) Argument could be raised to key in the numbers as they are and not mess around with them. *What is important is that you gain a feel for the numbers you enter and always evaluate your final answer.*

If you feel more comfortable with $0.1 \times 10^{-6} \times \dfrac{30000}{0.2} \times 100$, by all means do that.

2. This is another "taking control," or evaluation, statement. Whether you write them down or think about this kind of thing and *then* create an expression, you are shaping or controlling the outcome of the final answer.

3. 0.7 g enzymes in 100 g wet weight
   0.7 g enzymes in 90 g (= 90 mL) water

$$= \frac{0.7}{90} \text{ g/mL}$$

$$\text{Mass of enzymes in 1000 mL} = \frac{0.7}{90} \times 1000 \text{ g}$$

$$\text{mol/L of total enzymes} = \frac{0.7}{90} \times \frac{1000}{100\,000}$$

$$\text{mol/L of each enzyme} = \frac{0.7}{90} \times \frac{1000}{100\,000} \times \frac{1}{500}$$

$$\text{μmol/L of each enzyme} = \frac{0.7}{90} \times \frac{1000}{100\,000} \times \frac{1}{500} \times 10^6 \text{ (see comment below)}$$

$$= \mathbf{0.16\ μmol/L}$$

Comment: We *multiply* by 10^6 to convert mol to μmol.

4. (a) Consider 100 mL ethanol + 100 mL water

$$\text{Masses are 79 g + 100 g}$$

$$\% \text{ ethanol by weight} = \frac{79}{179} \times 100$$

$$= \mathbf{44.1\%}$$

(b) Consider 1000 mL of solution; this 1000 mL weighs 930 g

$$\text{Mass of ethanol in the 1000 mL} = 0.441 \times 930 \text{ g}$$

$$= 410 \text{ g}$$

$$\text{Number of moles of ethanol} = \frac{410}{46}$$

$$= 8.91 \text{ mol}$$

i.e., **8.91 mol/L**

5. $\text{mol} = \dfrac{40 \times 10^{-9}}{50\,000} = 8 \times 10^{-13} \text{ mol}$

$$= 0.8 \times 10^{-12} \text{ mol}$$

$$= \textbf{0.8 pmol}/100 \text{ mL}$$

$$= \textbf{0.08 pmol/L}$$

6. $0.8 \text{ }\mu\text{mol/L} = 0.8 \times 60\,000 \text{ }\mu\text{g/L}$

$$= 48\,000 \text{ }\mu\text{g/L}$$

$$= \textbf{48 }\mu\textbf{g/mL}$$

7. Final answers are shown in Table 2.5. Working and comments are shown below.
   (a) By inspection
   (b) $392.78 \times 10^{-2} \text{ g} = 3.9278 \text{ g} = \textbf{3.93 g}$
   (c) $448 \times 10^{-3} \text{ }\mu\text{mol} = 0.004\,48 \text{ }\mu\text{mol} = \textbf{4.48 nmol}$
   (d) $3.677 \times 10^{-5} \text{ }\mu\text{mol} = 36.8 \times 10^{-6} \text{ }\mu\text{mol} = \textbf{36.8 pmol}$

Note: The last step can be done by inspection, or you can take it back to the base unit and then recognise it as pico-.

$$36.8 \times 10^{-6} \times 10^{-6} \text{ mol} = 36.8 \times 10^{-12} \text{ mol} = 36.8 \text{ pmol}$$

Some advice: In the first step, convert the exponential notation to contain a "third" exponent ($^{-3}$, $^{-6}$, $^{-9}$, etc.), and then combine it with the prefixed unit. You will get something further down in the "multiples of three" set of numbers, which you will recognise by the prefix name. (Having written down the expression for the first step, you should be able to do the rest in your head.)

   (e) $1.904 \times 10^{-5} \text{ mol} = 19.0 \times 10^{-6} \text{ mol} = \textbf{19.0 }\mu\textbf{mol}$
   (f) $27.04 \times 10^{-4} \text{ nmol} = 2.704 \times 10^{-3} \text{ nmol} = \textbf{2.70 pmol}$

Table 2.5 Answers for Question 2.17	
(a) 0.0397 mmol/L	**39.7 μmol/L**
(b) 392.78×10^{-2} g	**3.93 g**
(c) 4.48×10^{-3} μmol	**4.48 nmol**
(d) 3.677×10^{-5} μmol	**36.8 pmol**
(e) 1.904×10^{-5} mol	**19.0 μmol**
(f) 27.04×10^{-4} nmol	**2.70 pmol**
(g) 380.4×10^{-1} nnol	**38.0 nmol**
(h) 4304×10^{-2} pmol	**43.0 pmol**

(g) 380.4×10^{-1} nmol = **38.0 nmol**

(h) 4304×10^{-2} pmol = **43.0 pmol**

8. The required details are entered in Table 2.6. Detailed calculations and comments are shown below.

(a) 90 000 g = 1 mol

$$80 \ \mu g = \frac{80 \times 10^{-6}}{90 \ 000} \times 1 \ mol$$

$$= 8.888 \times 10^{-10} \ mol$$

$$= 889 \times 10^{-12} \ mol$$

$$= \textbf{889 pmol}$$

Comment: With this type of calculation (i.e., one involving various submultiples, such as m, µ, n, p, etc.), it is best to work only in the "base units" and to do a final conversion once only at the end. Otherwise, you may end up going around in circles. Here, the µg was handled as g, and the other base unit mol was finally converted to pmol.

(b) $460 \ pg = \dfrac{460 \times 10^{-12}}{90 \ 000} \times 1 \ mol$

$$= 5.11 \times 10^{15} \ pmol$$

$$= \textbf{5.11 fmol}$$

(c) 1 mol = 90 000 g

$$80 \ mmol = \frac{80 \times 10^{-3}}{1} \times 90 \ 000 \ g$$

$$= 7 \ 200 \ g$$

$$= 7.20 \ kg$$

Comment: Yes, 7.20 kilograms! With that M_r, 1 mole is 90 kg. (This is what is meant by evaluating the answer.)

(d) $460 \ nmol = \dfrac{460 \times 10^{-9}}{1} \times 90 \ 000 \ g$

$$= 0.0414 \ g$$

$$= \textbf{41.4 mg}$$

(e) 80 µg/L = **889 pmol/L** (see part (a))

(f) $460 \ mg/L = \dfrac{460 \times 10^{-3}}{90000} \ mol/L$

$$= 5.11 \times 10^{-6} \ mol/L$$

$$= \textbf{5.11 µmol/L}$$

(g) $20 \text{ mL } 80 \text{ μmol/L} = \dfrac{20}{1000} \times 80 \times 90\ 000 \text{ μg}$

$$= 144\ 000 \text{ μg}$$

$$= \mathbf{144\ mg}$$

Comment: Here, everything was kept at the μ level. Seemed easier than going back to base units.

(h) $\dfrac{20}{1000} \times 460 \times 90\ 000 \text{ μg}$

$$= \mathbf{828\ mg}$$

Comment: Compare this answer to part (g) above.

Table 2.6   Answers for Question 2.18

	Enter (3 sig. figures)	Enter Prefix	unit
(a) 80 μg	889	p	mol
(b) 460 pg	5.11	f	mol
(c) 80 mmol	7.20	k	g
(d) 460 nmol	41.4	m	g
(e) 80 μg/L	889	p	mol/L
(f) 460 mg/L	5.11	μ	mol/L
(g) 20 mL 80 μmol/L	144	m	g
(h) 20 mL 460 μmol/L	828	m	g

3

Preparation of Solutions and Dilutions

As discussed in Chapter 2, concentrations of solutions may be expressed in gravimetric or chemical concentrations. Discussions on procedures for weighing and handling can be found in practical-oriented elementary chemistry books. The exercises here centre around the quantitative aspects. Students should be familiar with safety rules, such as always adding concentrated acids to water while stirring and not adding water to acid.

Practical and Commonsense Approaches Are Emphasised in These Exercises

1. Chemical balances weigh to 0.0001 g accuracy and the weighing unit is g (not mg or µg).
2. Simply because a solution is made in a standard or volumetric flask does not guarantee that it must be more accurate than a solution made in a measuring cylinder. Greater % errors are introduced in inaccuracies in initial weights or volumes than in inaccuracies in final volumes, as the exercises in this chapter will show.
3. Many solutions or liquids *can* be weighed on a balance and diluted according to weight. Do not, however, weigh concentrated acids releasing corrosive fumes *in* a chemical balance.
4. Much information is provided on the labels or packaging slips provided by chemical companies. These should be noted carefully not only for information that will be useful in calculations but also for safety and storage aspects.
5. **Approximately but exactly** is a term that students must understand and very often put into practice in preparing solutions. It means an exact value around a certain figure. For example, taking approximately 6 mL could mean taking an exactly known volume (e.g., 6.32 mL), and it is this exact volume that would be used in any calculation that followed. It does not mean "be sloppy"; it means do not waste time adjusting to 6.00 mL — as long as an exact ratio to something else is maintained. Another example where this practice would come in is in weighing materials. It may be impossible to weigh certain substances to exactly 1.0000 g, but any mass around one gram could be used (e.g., 0.9432 g), provided allowances are made in the final volume in which this mass is dissolved. A further situation in which this approach might apply concerns a final volume: it does not matter if it is not the exact figure you had in mind, provided it is around that figure. (It is the new figure that must be recorded and used in subsequent calculations.)
6. **Ratio Dilution Method:** A sensible approach to dilution would be 1.0 mL diluted to 346 mL rather than 1.445 085 705 mL diluted to exactly 500 mL. The first approach we will call the **Ratio Dilution Method;** the second usually involves applying the "formula

method." Note the interesting trick that can be played with the "Most Commonly Used Dilution Formula." Students would be familiar with:

$$C_1V_1 = C_2V_2 \text{ or } M_1V_1 = M_2V_2$$

C here refers to concentration, M to Molar concentration, and V to volume, of course. Watch what happens when we dilute a unit volume to a *final volume whose numerical value is the same as the starting concentration* — in other words, 1 mL 11.6 mol/L HCl when diluted to 11.6 mL. Why, the concentration becomes 1 mol/L, of course! The mathematical trick obviously is $1 \times 11.6 = 11.6 \times 1$. The volumes change in the same **ratio** as the concentrations but inversely. And, this trick really applies quite broadly: $5 \times 11.6 = 11.6 \times 5$ (5 mL diluted 11.6 mL becomes 5 mol/L).

This can come in very handy when we want a quick dilution that is very accurate, and we are dealing with volumes that are easily measured out accurately. The above dilution, 1 mL $\rightarrow$ 11.6 mL, also applies to 10 mL $\rightarrow$ 116 mL, 100 mL $\rightarrow$ 1160 mL, 50 mL $\rightarrow$ 580 mL, and so forth. (Please look at the answer to Question 3.9 for further advantages of this **Ratio Dilution Method.**)

Questions

1. How many grams of sucrose (M_r 342) would you need in order to make 500 mL of a 5 mmol/L solution?

2. Give the mass in grams of oestrogen (MW 270) required to make 200 mL of an 8.2 μmol/L solution.

3. How many grams of $K_3Fe(CN)_6$ (M_r 329) would be required to prepare 500 mL of a 50 mmol/L solution?

4. You are given a 2.27 g/L solution of glycine (M_r 75). If it were possible, how would you dilute this to give 100 mL of 17.5 mmol/L?

5. Assume a 36% (w/w) solution of protein (M_r 12 000) in water has a density of 1.09 g/mL. What volume would you need in order to prepare 500 mL 200 μg/mL solution?

6. Assume a 40% (w/w) solution of protein (M_r 9 000) in water has a density of 1.20 g/mL. What volume of this solution would you need in order to prepare 100 mL of 1 mmol/L protein solution?

7. What volume of concentrated HCl (28% w/w, specific gravity 1.15) would you dilute to 1.5 L to obtain a 0.6 mol/L solution of HCl? M_r HCl 36.5

8. What volume of saturated ammonium sulphate (SAS) must be added to:
   (a) 5 mL 50% SAS to make it 75%?
   (b) 10 mL 30% SAS to make it 57%?

   Assume that the volumes are additive.

9. You are provided with about 200 mL of a 1.74 mol/L solution of biotin (M_r 244). For an assay, you need a minimum of 100 mL of a 1 mmol/L solution. How would you perform the required dilution?

10. How would you dilute glacial acetic acid (17.4 mol/L) to give about 300 mL of 0.1 mol/L acetic acid?

11. How would you make a quick dilution of formic acid available as 23.6 mol/L to give *about* 200 mL of:
    (a) 1 mol/L?
    (b) 1 mmol/L?
    (c) 0.4 μmol/mL?

12. Commercial lactic acid (M_r 90) is available as an almost pure liquid (95% w/w). How would you prepare about 250 mL of a:
    (a) 2 mol/L solution?
    (b) 2 mmol/L solution?

13. Concentrated HCl is 11.6 mol/L. What volume would you dilute to *about* 200 mL to give a 10% w/v solution of HCl? M_r HCl 36.5

14. Concentrated sulphuric acid is 96% by weight H_2SO_4 (M_r 98) and has a density of 1.84 g/mL. Calculate the amount of acid required to make 1 L of 0.6 N H_2SO_4 (i.e., 0.6 Normal H_2SO_4).

15. What volume of a solution of $CaCl_2$, whose concentration is 2.0 mol/L and whose density is 1.084 g/mL, would be required to prepare:
    (a) 1.0 L of a 0.01 mol/L solution?
    (b) 400 mL of a 0.30 mol/L solution?

16. Commercially concentrated nitric acid is 69% by weight of HNO_3 (M_r 63) and its density is 1.41 g/mL. How would you prepare from it a solution that is 0.1 mol/L?

17. Assume a 40% (w/w) solution of protein (M_r 9 000) in water has a density of 1.20 g/mL. What volume of this solution would you need in order to prepare 100 mL of 400 μg/mL standard solution?

18. Glacial CH_3COOH (HAc) is 100% by weight and has a density of 1.05 g/mL. How would you prepare 600 mL of 0.5 mol/L HAc? M_r HAc 60

19. Commercial ammonia is available in two concentrations: *880 ammonia* and *910 ammonia*. These figures refer to their specific gravities: 0.880 and 0.910. Ammonia liquid is lighter than water, having a density of 0.677 at −34°C; and it boils at −33.4°C. When we dissolve it in water, we get ammonia solution. The more concentrated the ammonia, the lower the specific gravity; hence 880 ammonia is more concentrated than the 910. The 880 is 35% (w/w) and the 910 is 25% (w/w).
    (a) Calculate the molar concentration of the 880 ammonia (M_r 17).
    (b) How many mL of the 880 ammonia would you dilute to 1 L to give a 1 mol/L solution?

20. Using the 880 ammonia, how would you prepare 2 L of a chromatographic solvent: **n-butanol saturated with 4% ammonia.**

21. A stock reagent of the following description is required. Concentrations expressed are final concentrations. Carefully set out how you would prepare about 500 mL?
    - pH 7.0 Phosphate buffer 0.1 mol/L
    - Sucrose 5 mmol/L
    - Ag++ ions (as $AgCl_2$) 2 mmol/L

Information provided:

pH 7.0 Phosphate buffer is made by adding Na_2HPO_4 and NaH_2PO_4 of identical concentrations in the ratio of 3 volumes:2 volumes.

M_r Sucrose 342 A_r Na 23 H 1 P 31 O 16 Ag 108 Cl 35

22. A stock reagent of the following description is required. Concentrations expressed are final concentrations. Carefully set out how you would prepare 1 litre of the reagent.
    - pH 7 phosphate buffer 0.1 mol/L
    - Glucose 5 mmol/L
    - $CaCl_2$ 2 mmol/L
    - Protein 0.25 mmol/L

Information provided:

x mol/L pH 7 buffer is made by mixing x mol/L Na_2HPO_4 and x mol/L NaH_2PO_4 in the ratio 3 vol : 2 vol. Commercial protein (M_r 88 000) is available as a 17.6 % (w/v) solution.

M_r Glucose 180 A_r Na 23 H 1 P 31 O 16 Ca 40 Cl 35

Answers

1. Want 5 mmol/L

   $= 5 \times 342$ mg/L

   $= \dfrac{5 \times 324}{2}$ mg / 500 mL

   $= \mathbf{0.855\ g}$

   Note: Work in mg, as mmol quantities are required. Multiplying mmol by M_r gives the answer directly in mg. However, give the final answer in g, as you are expected to weigh the sucrose and make the solution and balances read in g.

2. 1 µmol/L = 270 µg/L

   $= \dfrac{270}{5}$ µg/200 mL

   8.2 µmol/L $= 8.2 \times \dfrac{270}{5}$ µg/200 mL

   $= 442.8\ µg^*$

   $= 0.4428$ mg

   $= \mathbf{0.000\ 443\ g}$

   Note: or directly 442.8 µg $\times 10^{-6} = 0.000\ 442\ 8$ g

This is too small a mass to weigh accurately on chemical balances. I would perhaps make the solution 1000 times as concentrated, and then dilute it 1000-fold. Weigh 0.4428 g → 1000 mL in a volumetric flask. Then 1 mL → 1000 mL volumetric flask. (Chemical laboratory rules must be followed. Dissolve mass in small volume, then make up to mark, mix, etc.)

3. 50 mmol/L = 25 mmol/500 mL

   mass in mg = 25×329

   $\qquad = 8225$ mg

   $\qquad = \mathbf{8.225\ g}$

4. 2.27 g/L = $\dfrac{2.27}{75}$ mol/L

   $\qquad = 0.03027$ mol/L $= 30.27$ mmol/L

Now that we have expressed the concentration in molarity, we can see it is possible to dilute the 30.27 mmol/L solution to a 17.5 mmol/L solution.

$M_1V_1 = M_2V_2$

$30.27\ V_1 = 17.5 \times 100$ mL

$\qquad V_1 = \dfrac{17.5 \times 100}{30.27}$

$\qquad\quad = 57.81$ mL

$\qquad\quad = \mathbf{57.8\ mL}$ (diluted to 100 mL)

5. Need 200 µg/mL

   $= 200 \times 500$ µg for 500 mL

   $= 100\ 000$ µg for 500 mL

   $= 0.1$ g / 500 mL

   $1.09 \times \dfrac{36}{100}$ g contained in 1 mL

   0.1 g contained in $\dfrac{0.1}{1.09} \times \dfrac{100}{36}$ mL

   $= \mathbf{0.25\ mL}$

6. 1 mmol/L contains 9 000 mg/L

   $= 900$ mg/100 mL

   $= 0.9$ g/100 mL

$$1.20 \times \frac{40}{100} \text{ g contained in 1 mL}$$

$$\Rightarrow 0.9 \text{ g contained in } \frac{0.9}{1.20} \times \frac{100}{40} \text{ mL}$$

= 1.875 mL

7. 1 g contains 0.28 g HCl

    1 mL weighs 1.15 g

    Need $1.5 \times 0.6 = 0.9$ mol HCl or 0.9×36.5 g HCl

    Mass of concentrated HCl needed $= 0.9 \times 36.5 \times \dfrac{100}{28}$ g

    Volume of concentrated HCl needed $= \dfrac{0.9 \times 36.5}{1.15} \times \dfrac{100}{28}$ mL

    = 102 mL

Note: The mL figure is *smaller* than the g figure because density (or specific gravity) is greater than 1. (If density is 0.9 g/mL, then more than 1 mL needs to be taken to have 1 g. If the density is 1.15 g/mL, then confidently reduce the mass by a ratio of $\frac{1}{1.15}$ to get the volume. This is the reason for dividing by 1.15 in the last step above.)

8. (a) Let x mL be the volume SAS to be added, and express percentages as decimals:

$$(5 \text{ mL})(0.50) + (x \text{ mL})(1.00) = 0.75(5 + x) \text{ mL}$$

$$2.5 + x = 3.75 + 0.75x$$

$$0.25x = 1.25$$

$$x = \textbf{5 mL}$$

Note: The method of calculation shown here is the **full calculation method** and should be used for tricky volumes and concentrations. Students can see by inspection that mixing equal volumes of 50% SAS and 100% SAS will give 75% SAS.

(b) The more tricky one:

$$(10 \text{ mL})(0.30) + (x \text{ mL})(1.00) = 0.57(10 + x) \text{ mL}$$

$$3 + x = 5.7 + 0.57x$$

$$0.43x = 2.7$$

$$x = \frac{2.7}{0.43}$$

$$= \textbf{6.28 mL}$$

9. By **ratio dilution method:**

1.74 mol/L = 1740 mmol/L

1 mL → 1740 mL = 1 mmol/L

0.1 mL → 174 mL = 1 mmol/L

Note: → symbol is used to show that 1 mL is diluted to a final volume of 1740 mL.

Let's compare the **ratio dilution method** to the commonly used **formula method:**

$$M_1V_1 = M_2V_2$$

$$1740V_1 = 1 \times 100$$

$$V_1 = \frac{1 \times 100}{1740}$$

$$= 0.057\ 471\ 263\ 678\ 160\ 9 \ldots \text{mL}$$

This volume has to be made up to 100 mL!

Clearly, the ratio dilution method is the sensible way to perform this dilution. It is a quick, yet very accurate method. Note that 0.1 mL is easy to measure accurately. It is impossible to measure out the volume required in the formula method. (In the interest of fair play, let's convert 0.057 471 263 678 160 9 ... mL, to the usual number of significant figures. 0.0575 mL (or even 0.058) is still impossible to measure out!) Also, consider the magnitude of errors that could be introduced. Suppose, there was an error of 0.005 mL in the initial volume of the formula method. The error would be 0.005/0.05747 = 8.7%. Suppose the final volume in the ratio method was out by a whole 2 mL: the error would be only 2/174 = 1.1%. Even at 5 mL, the error would be 5/172 = 2.9%. The reader will readily agree that one drop more or less of the concentrate when small volumes are involved will cause a greater percentage error than discrepancies at the other end (the final volume). Note in the ratio method that the more exacting glassware (pipette) and round figure volumes are used at the difficult end. Also, for our calculations, which always must include evaluation, *the ratio approach is easier to visualise.*

However, consider each situation in terms of practicality, and take into account the volumes involved and the accuracies of glassware used (and toxicity and toxic vapours). For measuring larger round number volumes (100 mL, 500 mL, etc.), standard flasks are more accurate than measuring cylinders and you can more easily adjust to the graduation mark in the former.

10. By **ratio dilution method:**

1 mL 17.4 mol/L → 17.4 mL = 1 mol/L

1 mL 17.4 mol/L → 174 mL = 0.1 mol/L

2 mL 17.4 mol/L → 348 mL = 0.1 mol/L

By **formula method:**

$$C_1V_1 = C_2V_2$$

$$17.4\ V_1 = 0.1 \times 300$$

$$V_1 = \frac{0.1 \times 300}{17.4}$$

$$= \textbf{1.724\ 138\ mL}\ \text{(to be diluted to 300 mL)}$$

Comment: The expected answer is the one given by the ratio dilution method.

11. (a) 1 mL $\rightarrow$ 23.6 mL = 1 mol/L

Or **10 mL $\rightarrow$ 236 mL = 1 mol/L**

That is, simply dilute 10 mL to 236 mL in a measuring cylinder.

(b) 1 mL $\rightarrow$ 23.6 mL = 1 mol/L

Or 1 mL $\rightarrow$ 23 600 mL = 1 mmol/L

This is the correct concentration of solution (that we want), but there are two problems:

(i) We have wasted one whole mL (need only much less). Wastage can be a concern when the material is expensive (e.g., an enzyme or antibody) or when we have precious little of a preparation.

(ii) We have to handle such a large volume: 23 600 mL (23.6 L)!

In most situations, the second problem is the more serious one. Therefore, a **two-step dilution** approach is necessary:

0.1 mL $\rightarrow$ 236 mL = 0.01 mol/L or 10 mmol/L

then

20 mL of the 10 mmol/L $\rightarrow$ 200 mL = 1 mmol/L

(c) 0.4 μmol/mL = 0.4 mmol/L

As in part (b)

0.1 mL $\rightarrow$ 236 mL = 10 mM

then

0.4 mL $\rightarrow$ 10 mL = 0.4 mmol/L

or 8 mL $\rightarrow$ 200 mL

or **10 mL $\rightarrow$ 250 mL**

12. (a) Require $\dfrac{250}{1000} \times 2$ moles

$= \dfrac{250}{1000} \times 2 \times 90$ g if 100%

$= \dfrac{250}{1000} \times 2 \times 90 \times \dfrac{100}{95}$ g if 95%

= 47.3684 g

Weigh this amount as accurately as possible and dissolve then dilute to 250 mL. (See comments following part b.)

(b) It may be wasteful to make the 2 mol/L and then dilute 1:1000 or 0.25 mL to 250 mL (to give the required 2 mmol/L). Furthermore, there is no indication that part (b) is meant to be a follow-on from part (a). If we are to approach this part independently, then:

Weigh a thousandth of the above (i.e., **0.0474 g**), dissolve, and dilute to **250 mL**.

It would be difficult to control the drop size of this syrup to get exactly the above weight; therefore, take any mass close to 0.0474 and dilute to an appropriate volume. The ratio to be followed is:

0.0474 : 250

Therefore, a mass of **0.0626 g** would be diluted to:

$\dfrac{0.0626}{0.0424} \times 250 =$ **330 mL.**

Note: This is an example of the "approximately but exactly" method discussed in the introduction to this chapter. This approach may also be necessary for part (a).

13. By **ratio dilution method:**

11.6 mol/L = 11.6 × 36.5 g/L

$= \dfrac{11.6 \times 36.5}{10}$ g/100 mL

= 42.34 % w/v

Have 42.34%; want 10%.

10 mL 42.34% diluted to 42.34 mL = 10%

or **50 mL 42.34% diluted to 212 mL** = 10%

Note: The last line is obtained by multiplying both 10 and 42.34 by 5.

By **formula method:**

$$C_1 V_1 = C_2 V_2$$

$$42.34 \; V_1 = 10 \times 200$$

$$V_1 = \frac{10 \times 200}{42.34}$$

$$= 47.2367 \text{ mL} \rightarrow 200 \text{ mL}$$

Note: The ratio dilution method is the preferred answer.

14. $M_r \; H_2SO_4 = 98$

Equivalent weight $= 49$

Require 0.6×49 g/L

1 mL concentrated H_2SO_4 weighs 1.84 g

1 mL concentrated H_2SO_4 contains 1.84×0.96 g H_2SO_4

i.e., 1.84×0.96 g in 1 mL

0.6×49 g in $\dfrac{0.6 \times 49}{1.84 \times 0.96}$ mL

$$= \mathbf{16.64 \; mL}$$

15. The question simply is this: If you are given a 2 mol/L solution, how could you prepare:
    (a) 0.01 mol/L?
    (b) 0.30 mol/L?

The density does not come into the calculations.

(a) 1 mL 2 mol/L $\rightarrow$ 200 mL = 0.01 mol/L

   or **5 mL** 2 mol/L $\rightarrow$ 1 L = 0.01 mol/L

(b) 3 mL 2 mol/L $\rightarrow$ 20 mL = 0.3 mol/L*

   or **60 mL** 2 mol/L $\rightarrow$ 400 mL = 0.3 mol/L

*Note: The $M_1 V_1 = M_2 V_2$ is used in figuring this line. With experience, this line can be visualised immediately. The products on both sides of the equation are equal. The figure that is volume on one side becomes the molarity on the other (including shifting the decimal point as needed).

$3 \times 2 = 20 \times 0.3$

$\uparrow$ This figure is 0.3 instead of 3, as we need 0.3 mol/L.

Formula method ($M_1V_1 = M_2V_2$) may be used as well.

For (b), $2V_1 = 0.30 \times 400$

$$V_1 = \frac{0.30 \times 400}{2}$$

$$= \textbf{60 mL}$$

16. Calculation-wise it is simpler to prepare by weight, but you would not weigh concentrated HNO_3 with corrosive and toxic vapours *in* a chemical balance, would you? Maybe on a top-loading balance. Be careful.

If **by weight,** density does not come into the calculations.

M_r HNO_3 63

Need 6.3 g/L for a 0.1 mol/L   (i)

Since commercial HNO_3 is only 69%

$6.3 \times \dfrac{100}{69} = \textbf{9.13 g/L}$ is required   (ii)

By volume, $6.3 \times \dfrac{100}{69} \times \dfrac{1}{1.41} = \textbf{6.48 mL/L}$ (iii)

Comment: Check and evaluate all the time. (ii) is greater than (i) because only 69%. (iii) is less than (ii) because density is greater than 1. Concentrated HNO_3 is a heavy liquid; 6.48 mL weighs 9.13 g. You can check this: $\frac{9.13}{6.48} = 1.41$ (the density given). (If the occasion presents itself, this would be a brilliant way to check/evaluate your answers.)

17. 400 µg/mL = 40 000 µg/100 mL = 40 mg/100 mL = 0.04 g/100 mL

$1.2 \times \dfrac{40}{100}$ g contained in 1 mL

$\Rightarrow$ 0.04 g contained in $\dfrac{0.04}{1.2} \times \dfrac{100}{40} = \textbf{0.083 mL}$

18. Need $60 \times 0.5 \times 0.6$ g HAc = 18 g

There are 1.05 g in 1 mL

18 g in $\dfrac{18}{1.05} \times 1$ mL

$= \textbf{17.14 mL}$

19. (a) 0.88 g is the mass of 1 mL of solution

100 g is the mass of $\dfrac{100}{0.88}$ = 113.6 mL of solution

113.6 mL contains 35 g NH_3

113.6 mL contains $\dfrac{35}{17}$ mol NH_3

1 L contains $\dfrac{1000}{113.6} \times \dfrac{35}{17}$ mol NH_3

= 18.1 mol

Molarity = 18.1 mol/L

(b) $M_1V_1 = M_2V_2$

$(18.1) V_1 = 1.0 \times 1.0$

$V_1 = \dfrac{1}{18.1}$

     = 0.0552 L

     = 55.2 mL

20. Use the information given in Question 19.

Dilute 4 volumes of 35% $\rightarrow$ 35 volumes.

This would give a 4% solution.

(The actual volumes could be 40 mL to 350 mL.)

Add this 4% ammonia solution to 2 L n-butanol, mix thoroughly, allow to settle, and make certain that two layers appear; use the top layer. If two layers do not appear, add more 4% ammonia.

Note: While the question does not say *about* 2 L, the volume produced here, an excess of 2 L, is exactly what would be expected in this *practical* situation. Also, there could be arguments as to what is 4%. Ammonia poses a number of problems: it is a gas; in liquid state, it is lighter than water; and when it dissolves in water, part of it reacts with water to become NH_4OH. So, besides the % (w/w) or (w/v) problem, we are faced with all these other problems. Irrespective of whatever else is happening, the chromatographic solvent is meant to be as prepared as above — after all, there is "trial and error" involved in "if two layers do not appear, add more 4% ammonia."

21. M_r $Na_2HPO_4 = 142$ 14.2 g/L $= 0.1$mol/L or **4.26 g/300 mL (A)**

M_r $NaH_2PO_4 = 120$ 12.0 g/L $= 0.1$mol/L or **2.4 g/200 mL (B)**

Sucrose 342×5 mg/L $= 5$ mmol/L

$= 1710$ mg/L $= 855$ mg / 500 mL $= $ **0.855 g/500 mL**

M_r $AgCl_2 = 178$

$178 \times 2 = 356$ mg/L $= 2$ mmol/L or 178 mg/500 mL

$$= \text{0.178 g/500 mL}$$

Method: Mix A and B, and then dissolve the mixture in 0.855 g sucrose and 0.178 g $AgCl_2$.

22. In calculations, cater for:

600 mL Na_2HPO_4 (M_r 142) $= 142 \times 0.6 \times 0.1 = $ **8.52 g (A)**

400 mL NaH_2PO_4 (M_r 120) $= 120 \times 0.4 \times 0.1 = $ **4.8 g (B)**

Do not make the 600 mL and 400 mL; we need to leave space for the protein, which is in an aqueous solution.

For 5 mmol/L glucose, we need $180 \times 5 \times 10^{-3}$ g $= $ **0.90 g/L (C)**

For 2 mmol/L $CaCl_2$ (M_r 110), we need $110 \times 2 \times 10^{-3} = $ **0.22 g/L (D)**

For 0.25 mmol/L protein, we need $88\,000 \times 0.25 \times 10^{-3} = 22$ g

22 g is contained $22 \times \dfrac{100}{17.6}$ mL $= $ **125 mL (E)**

Note: What we have done here is to take sufficient protein so that when it is diluted as a result of adding it to the buffer, it will attain the correct final concentration.

Method: Dissolve A, B, C, and D (solids) in about 800 mL distilled water in a beaker. Transfer to a 1 L volumetric flask or measuring cylinder. Now add the protein solution (E) and make up to 1 L with distilled water.

Note: It does not matter when the glucose and $CaCl_2$ are added: as above, or after one or both Na_2HPO_4 and NaH_2PO_4 are added and made up to about 800 mL. No harm will come to the glucose and $CaCl_2$, **but** the protein must only be added **after** the buffer is made up (to pH 7). Exposing the protein to either HPO_4^{2-} or $H_2PO_4^-$ before the pH 7 buffer is established could denature the protein. As a general rule, whether in preparation of a stock or in experimental set-up, *always add biochemicals (that are liable to denaturation) last.* The commonly encountered biochemicals that are liable to denaturation in undergraduate biochemistry courses are proteins and nucleic acids. Remember also, when an addition is in a liquid or solution form, allowances have to be made to accommodate the volumes involved.

4

Relative Molecular Mass and Stoichiometry

The chemical formula of a compound tells us the number and the kind of atoms that make up a molecule of the compound. Adding the relative atomic masses (A_r) gives us the relative molecular masses (M_r) of molecules. The fact that whole atoms are found in molecules and consequently replaced or released in reactions means that this phenomenon is apparent in virtually all aspects of quantitative chemistry. *Stoichiometry* is the branch of chemistry that deals with the quantitative relationships of combining elements. Stoichiometric calculations involve the chemical composition or atomic make-up of molecules, relationships in chemical reactions (equations) and composition of solutions. This chapter deals with chemical composition and reactions.

The elementary concepts and principles applying to simple molecules and studied in junior chemistry courses also apply to complex biochemical molecules and reactions.

1. **Percentage composition** is the percent by weight of each type of atom in the compound. For example, from the formula H_2O, the percentages by weight of H and O can be calculated as 11% and 89%, respectively. Conversely, if the percentages by weight are known, the formula can be worked out.

2. A formula computed from the percentage composition only gives the simple ratio of the atoms involved. Such a formula is called the **empirical formula** and may not be the **true** or **molecular formula** (e.g., the empirical formula of benzene is CH and glucose is CH_2O, whereas their molecular formulae are C_6H_6 and $C_6H_{12}O_6$, respectively). Empirical formula weight is often the minimum formula weight. Additional information besides percentage composition is necessary for working out a molecular formula and M_r. M_r is the sum of the A_r in the molecular formula.

3. In high school chemistry, the knowledge of percentage composition basically enabled the elucidation of formulae of inorganic and simple organic compounds, including biochemical monomers. A feature of biochemical stoichiometry is the extension of these principles to macromolecules, such as carbohydrates, proteins, and nucleic acids. In place of atomic or elementary composition, we may deal with amino acid or nucleotide composition in gaining an understanding of the formulae or structural make-up of the biochemical polymers. (**Note:** Certain aspects of biochemical stoichiometry are continued in Chapter 13, "DNA and Molecular Biology.")

Questions

---

1. Calculate the percentages of hydrogen and oxygen in water (H_2O).
   A_r H 1 O 16

2. Given that 0.160 g copper when heated in oxygen becomes copper oxide of mass 0.200 g, calculate the formula for copper oxide.
   A_r O 16 Cu 63.6

3. The percentage composition of penicillin is:
   C 56.1% H 7.6% N 8.2% S 9.4% O 18.7%
   Determine its empirical formula.
   A_r C 12 H 1 N 14 S 32 O 16

4. What empirical formula would be produced from the following percentage compositions? What are the most likely molecular formulae, and what are the compounds?
   (a) C 92.4% H 7.6%
   (b) C 75.0% H 25.0%
   A_r C 12 H 1

5. Given M_r glucose is 180 and A_r H 1 C 12 O 16, calculate M_r of the following by comparing the structures to glucose or the other molecules in the list:
   • Ribose
   • 2-deoxyribose
   • Rhamnose (6-deoxymannose)
   • Fructose
   • Sucrose
   • Raffinose (α-galactosylsucrose)

6. The oxygen content of serine is 45.71% and the nitrogen content 13.33%. Calculate the minimum or empirical molecular weight of serine based on each of the percentage compositions. Which is the more likely true molecular weight? A_r O 16 N 14

7. 0.87 mmol of acid (H^+) was taken up by 1 g protein. (This is due to the basic groups becoming protonated.) Calculate the minimum M_r. Suppose it is known that these are 42 basic groups, estimate the true M_r.

8. It is known that two phosphates are needed in the activation of each molecule of muscle glycogen phosphorylase by adenosine triphosphate (ATP). In an incubation, ATP and phosphorylase reacted at the rate of 1 μmol ATP to 94.5 mg phosphorylase. Calculate the minimum and true M_r of phosphorylase.
   M_r ATP 507 ADP 427 PO_4^{3-} 95

9. A molecule of a respiratory protein is known to contain two copper-bound non-amino acid residues. An analysis of this protein gave the following results:
   • Cu (A_r 63.5) 0.34%
   • Amino acids residues 98.24%

   Calculate the M_r of the organic residue.

10. Estimate the molecular weight of a globular protein composed of 100 amino acids.

11. 0.71 g of a peptide (M_r 1050 ± 10) was hydrolysed and the amino acids estimated as gly 0.29 g, ala 0.18 g, and phe 0.35 g.
M_r gly 75 ala 89 phe 165
(a) Calculate the % increase of mass upon hydrolysis. Explain why this would occur.
(b) Determine the formula for the peptide; express the formula as gly_x ala_y phe_z.

12. Haemoglobin (Hb) contains 0.345% iron. Calculate the minimum relative molecular mass.
A_r Fe 56

13. 3.46 g protein was hydrolysed and its tryptophan (M_r 204) content was determined to be 3.1 mg.
(a) Calculate the minimum relative molecular mass of the protein.
(b) If there are four residues of tryptophan per molecule of the protein, calculate the true M_r.

14. A protein known to contain one Fe per molecule is 0.20% iron and 6.12% arginine. Calculate the number of arginine residues per molecule.
A_r Fe 56 M_r arg 174

15. 35.0 mL 0.20 mol/L HCl required 17.5 mL NaOH for titration using methyl orange as indicator. What is the concentration of NaOH?

16. Given:

$$ATP \longrightarrow ADP + P_i$$
$$M_r\ 507 \qquad 427 \quad P_i\ (or\ PO_4^{3-}) = 95$$

Explain why there is a discrepancy when the masses on the LHS and RHS of the equation are compared.

17. Some aspects of photosynthesis can be demonstrated with isolated chloroplasts if an electron acceptor for the photolysis of water is provided. Ferricyanide can act as the electron acceptor:

$$4Fe(CN)_6^{3-} + H_2O \longrightarrow 4Fe(CN)_6^{4-} + 4H^+ + O_2$$

In this experiment it is appropriate to use a 50 mmol/L solution of potassium ferricyanide. Assuming the reaction proceeds completely to the right, calculate the volume of 50 mmol/L $K_3Fe(CN)_6$ required to permit the release of 0.50 μmol oxygen.

18. Yeast ferments glucose, yielding ethanol and carbon dioxide according to:

$$C_6H_{12}O_6 \longrightarrow 2C_2H_5OH + 2CO_2 + Energy$$

In an experiment, yeast, in 180 mL 10% (w/v) glucose solution, gave off 2 L of carbon dioxide. Determine the concentration of glucose at the end of the experiment. Give answer as % w/v.

Assume that the experiment was performed at normal or standard conditions of temperature and pressure (i.e., NTP or STP). One mole of any gas at NTP occupies a volume of 22.4 L. A_r C 12 O 16 H 1

Answers

1. 1 mole H_2O = 2 + 16 = 18 g

$$\% \text{ H} = \frac{2}{18} \times 100 = \mathbf{11.1\%}$$

$$\% \text{ O} = \frac{16}{18} \times 100 = \mathbf{88.9\%}$$

2. 0.200 g copper oxide is made up of 0.160 g Cu and 0.040 g oxygen.

$$\text{Moles}^* \text{ of Cu} = \frac{0.160}{63.6} = 0.00251$$

$$\text{Moles of O} = \frac{0.040}{16} = 0.00250$$

The ratio of the two is 1:1.

Therefore, copper oxide is Cu_1O_1 or **CuO**.

*****Note:** The older term **gram-atoms** was used instead of moles when referring to elements or atoms.

3.

Table 4.1   Answer to Question 4.3

Element	C	H	N	S	O
Atomic ratios	$\frac{56.1}{12}$	$\frac{7.6}{1}$	$\frac{8.2}{14}$	$\frac{9.4}{32}$	$\frac{18.7}{16}$
=	4.675	7.6	0.586	0.294	1.169
Divide by 0.294 (to get ratios)	15.90	25.85	1.99	1.00	3.98
Integers	16	26	2	1	4

Empirical formula $\mathbf{C_{16}H_{26}O_4N_2S}$

Note: After dividing by A_r, do not round-off too much; leave two or three decimal places — we need the accuracy as there is another division (by the smallest number) to follow. *Rounding off too much too soon in this and in most situations must be avoided.*

It is chemical convention to write organic compounds as C H O followed by N and S.

4. (a)

Table 4.2   Answer to Question 4.4(a)

Element	C	H
Atomic ratios	$\frac{92.4}{12}$	$\frac{7.6}{1}$
=	7.7	7.6

As the ratio is 1:1, empirical formula is **CH**. The only way carbon and hydrogen can combine at a ratio of 1:1 is as $\mathbf{C_6H_6}$, which is **benzene**. (C_6H_6 is the molecular formula.)

(b)

Element	C	H
	Table 4.3 Answer to Question 4(b)	

Element	C	H
	$\dfrac{75}{12}$	$\dfrac{25}{1}$
=	6.25	25
Ratios	$\dfrac{6.25}{6.25}$	$\dfrac{25}{6.25}$
=	1	4

Empirical formula is CH_4, which has to be the molecular formula. We cannot have any value for n in the expression $n(CH_4)$ other than $n = 1$ and satisfy the bonding between C and H. So the empirical formula and the molecular formula are the same. CH_4 is **methane**.

5. Ribose, a pentose, is one CHOH less than glucose:

$180 - (12 + 18) = \mathbf{150}$

2-deoxyribose has an H instead of an OH at C-2 (compared to ribose):

$150 - 17 + 1 = \mathbf{134}$

Rhamnose has a CH_3 at C-6 instead of a CH_2OH (compared to glucose):

$180 - 31 + 15 = \mathbf{164}$

Fructose is an isomer of glucose:

$M_r = \mathbf{180}$

Sucrose is a disaccharide; one H_2O is lost when glucose and fructose combine:

$180 \times 2 - 18 = \mathbf{342}$

Raffinose is a galactose-containing trisaccharide; two H_2O are lost:

$3(180) - 2(18) = \mathbf{504}$

Note: The purpose of this question is to encourage tertiary students to use authority and commonsense in working out or visualising quantitative features relating to structures with which the students are familiar; there is no need to go back to a junior high school approach of listing all the atoms and adding up the A_r. This approach also helps to reinforce understanding of the structures of these molecules, and what happens to them reaction-wise.

6. Based on O, minimum $M_r = \dfrac{16}{45.71} \times 100 = 35$

Based on N, minimum $M_r = \dfrac{14}{13.33} \times 100 = 105$

Note: Two different minimum M_r have been obtained. The O-based minimum M_r is smaller than the N-based minimum M_r; this is because there are more O atoms in a serine molecule than there are N atoms. Even if we did not know the formula of serine, such a deduction (smaller minimum M_r relating to more atoms) is generally true. From the experimental results, all we can conclude is that the true M_r is either **105** or a multiple of 105. (The fact that the O-based minimum M_r is exactly one-third of the N-based minimum M_r tells us that there are three oxygen atoms to every one nitrogen atom.) However, we need one further piece of information — that there is only one N atom in the molecule — before we can conclude that the true M_r is 105. (We now know that there is indeed one N atom in serine and the true M_r is 105.)

7. 0.87 mmol combines with 1000 mg protein.

1 mmol combines with $\dfrac{1000}{0.87}$ = 1149 mg.

Minimum M_r = 1149.

This is based on the assumption that there must be at least one acid-binding group in the protein molecule.

Since we are told that there are 42, the true M_r = 1149 × 42 = 48 258, or 48 000.

8. 1 mole ATP releases 1 mole PO_4^{3-}, which is taken up by the phosphorylase.

If 1 μmole ATP reacts with 94.5 mg or 94 500 μg, then 94 500 is the minimum M_r.

The information provided says two phosphates:

$\Rightarrow M_r = 2 \times 94\ 500 = \mathbf{189\ 000}$

9. % organic residue = 100 − 98.24 − 0.34

$$= 1.42$$
$$0.34\% = 2 \times 63.5$$
$$\Rightarrow 1.42\% = \frac{1.42}{0.34} \times 2 \times 63.5$$
$$= 530.4$$

530.4 represents two organic residue

$\Rightarrow M_r$ of each is **265**

10. The average M_r of an amino acid is about 128. The smallest glycine is 75.

The largest tryptophan is 204.

Generally, the smaller ones predominate in most proteins.

The weighted average is conveniently taken as 128.

When amino acids combine, one H_2O (M_r 18) is lost.

This accounts for the average residue weight being 128 − 18 = 110.

Hence, the globular protein with 100 amino acids should have an estimated M_r of 100 × 110 = **11 000**.

Note: Some estimated relative molecular masses and actual (or best experimentally obtained) relative molecular masses for a few proteins are shown in Table 4.4.

Table 4.4   True and Estimated Relative Molecular Masses of Proteins

Protein	Amino acid residues	Estimated M_r	Actual M_r
Insulin (bovine)	51	5 610	5 733
Ribonuclease A (bovine pancreas)	124	13 640	13 700
Lysozyme (egg white)	129	14 190	14 000
Human haemoglobin (including 4 haems)	574	65 606	64 500
Human haemoglobin (without 4 haems)	574	63 140	62 000

11. (a) Mass of amino acids = 0.29 + 0.18 + 0.35
$$= 0.82 \text{ g}$$
Original peptide = 0.71 g
Increase = 0.11 g
$$\% \text{ increase} = \frac{0.11}{0.71} \times 100$$
$$= \mathbf{15.5\%}$$

The increase is due to H_2O joining at the hydrolysed sites.

Note: The amino acids resulting from protein hydrolysis are supposed to have an average M_r of 128, and the amino acid residues 110. If these figures are also assumed for this peptide, then the increase in weight due to hydrolysis should be:

$$\frac{18}{110} \times 100 = 16\%$$

(b) Moles of amino acids in the hydrolysate:

gly $\frac{0.29}{75} = 0.0039$

ala $\frac{0.18}{89} = 0.0020$

phe $\frac{0.35}{165} = 0.0021$

Ratios = 2:1:1
Hence empirical formula is gly_2 ala_1 phe_1.
The formula weight for this formula is
$(2 \times 75 + 89 + 165) - 3H_2O$ (or 3×18) = 350
Ratio between FW and true M_r is:
$$\frac{1050}{350} = 3$$
Hence, the true formula is $\mathbf{gly_6 \ ala_3 \ phe_3}$.

12. At least one atom of Fe must be present per molecule of Hb. Therefore, the minimum molecular weight is that mass of protein in which 56 g represents 0.345%.

$0.345\% = 56$
$$100\% = \frac{100}{0.345} \times 56 = \mathbf{16\ 232}$$

Note: The quaternary structured Hb contains four Fe atoms, giving a true M_r of:

$$4 \times 16\ 232 \approx 64\ 500.$$

13. (a) M_r trp = 204, residue weight = 204 − 18 = 186
3.1 mg represents 186
$$\Rightarrow 3460 \text{ mg represents } \frac{3460}{3.1} \times 186$$
$$= \mathbf{207\ 600} \text{ (minimum } M_r)$$

(b) True M_r = 4 × 207 600
$$= \mathbf{830\ 400}$$

14. $0.2\% = 56$

$100\% = \dfrac{100}{0.2} \times 56 = 28\,000$ (this is the M_r of the protein)

6.12% of $28\,000 = 1713.6$

Each arg residue $= 174 - 18 = 156$

Number of arg residues $= \dfrac{1713.6}{156}$

$= 10.98$

$= \mathbf{11}$

15. $HCl + NaOH \longrightarrow NaCl + H_2O$

Only half the volume of NaOH is required. So the concentration must be twice as much (i.e., **0.40 mol/L**).

Note: There is no need to show calculations when chemical logic is clearly stated or is obvious. This question is thrown in here just to keep your *basic* confidence going. If you figured this out just by looking at it, hang on to that confidence. You will do *all* your biochemical calculations almost just as easily.

16. The given equation is only a biochemist's short-form version. The full reaction is

$$ATP^{4-} + H_2O \longrightarrow ADP^{3-} + HPO_4^{2-} + H^+$$
507   18   427   96   1

This question is provided as a warning to students that, in stoichiometric calculations, full (and balanced) equations must be used. However, biochemical abbreviations such as ATP and $NADP^+$ and equations such as the one in the question and $NAD^+ \longrightarrow NADH + H^+$ are meant to be used by biochemistry students in all other work; make certain that proper conventions (which have been based on sound reasons) are followed (e.g., positive sign in NAD^+).

17. $4Fe(CN)_6^{3-} + H_2O \longrightarrow 4Fe(CN)_6^{4-} + 4H^+ + O_2$

4 mol   release   1 mol

2 μmol   release   0.5 μmol

What volume 50 mmol/L $K_3Fe(CN)_6$ would contain 2 μmol?

50 μmol contained in 1 mL

2 μmol contained in $\dfrac{2}{50} \times 1 = \mathbf{0.04\ mL\ or\ 40\ \mu L}$

18. $C_6H_{12}O_6 \longrightarrow 2C_2H_5OH + 2CO_2 + E$

1 mol (180 g)   produces   44.8 L

44.8 L is produced by 180 g

$\Rightarrow$ 2 L is produced by $\dfrac{2}{44.8} \times 180 = 8.2$ g

Original glucose $= 18.0$ g

Glucose used up $= 8.2$ g

Remainder $= 9.8$ g

Concentration $= \dfrac{9.8}{180} \times 100\%$ w/v

$= \mathbf{5.5\%\ w/v}$

5

Acids, Bases, and Buffers

In biochemistry, the most useful definition of an **acid** and a **base** is that of Brönsted. Acids are substances that donate hydrogen ions, and bases are substances that accept hydrogen ions. Table 5.1 lists some common inorganic, organic, and biochemical compounds that can ionise to give hydrogen ions, and their corresponding **conjugate bases.**

Hydrogen ions (H^+) exist in water as hydronium ion (H_3O^+).

$$H^+ + H_2O \rightarrow H_3O^+$$

In simple terms, a hydrogen ion is the same as a proton. The terms H^+, proton, and H_3O^+ are generally interchangeable.

The Brönsted definitions, particularly the concept of a conjugate base, do cause some confusion. Note in Table 5.1 that NH_4^+ is denoted as an acid. We all know that NH_4OH, like NaOH, is a base or an **alkali** (an alkali is an aqueous solution of a base). Similarly, $-COOH$ is an acid group, whereas $-NH_2$ is a basic group. What about $-NH_3^+$?

Most of the difficulties can be resolved by considering:

(a) the extent of dissociation or ionisation — it is this concept that introduces terms such as **strong and weak acids and bases.**
(b) whether the equilibrium of an ionisation, or the reaction with water, is to the right or left.

In the Brönsted definition, water is a base!

$$H^+ + H_2O \rightarrow H_3O^+$$

The most important practical definitions of all these **electrolytes** (the collective term for acids, bases, and salts) is what they do with respect to water. Water dissociates to a very small extent.

$$H_2O \rightleftharpoons H^+ + OH^-$$

or as chemists prefer it:

$$H_2O + H_2O \rightleftharpoons H_3O^+ + OH^-$$

The molar concentration of water is 55.5 mol/L and of H_3O^+ and OH^- is 10^{-7} mol/L each. It is this feature that gives us the pH scale —the pH of water and *neutral* solutions is 7, *acidic* solutions have pH less than 7, and *basic* solutions have pH greater than 7. Acids then are substances that increase

Table 5.1   Brönsted Acids and Their Corresponding Conjugate Bases

Brönsted Acid		Conjugate Base		Proton
HCl	$\rightarrow$	Cl$^-$	+	H$^+$
CH$_3$COOH	$\rightarrow$	CH$_3$COO$^-$	+	H$^+$
NH$_4^+$	$\rightarrow$	NH$_3$	+	H$^+$
H$_3$PO$_4$	$\rightarrow$	H$_2$PO$_4^-$	+	H$^+$
H$_2$PO$_4^-$	$\rightarrow$	HPO$_4^{2-}$	+	H$^+$
COOH | CH$_2$ | NH$_2$ Glycine	$\rightarrow$	COO$^-$ | CH$_2$ | NH$_2$ Glycine	+	H$^+$
COOH | CH$_2$ | NH$_3^+$ Glycine	$\rightarrow$	COOH | CH$_2$ | NH$_2$ Glycine	+	H$^+$

the hydrogen ion concentration (of water) to greater than 10^{-7} mol/L and, conversely, bases decrease the hydrogen ion concentration to less than 10^{-7} mol/L.

The pH Scale

Chemists have devised a simple way of writing numbers associated with low concentrations, such as 4×10^{-11} mol/L or 2×10^{-3} mol/L. They have defined a number called the pH. Mathematically:

$$pH = -\log_{10} [H^+]$$

Indeed, very cleverly, the p-scale of numbers is used also in terms such as pK (negative log of dissociation constant) and pI (isoelectric point). This consistency of the use the p-scale makes it easy to explain the ionisation of organic groups and the nature of charges on molecules.

If we look at the dissociation of water again:

$$H_2O \rightleftharpoons H^+ + OH^-$$

$$K = \frac{[H^+][OH^-]}{[H_2O]}$$

As there is negligible loss of the 55.5 mol/L H_2O in producing the 10^{-7} mol/L H^+ and OH^-, a new constant may be obtained:

$$K[H_2O] = [H^+][OH^-]$$

or

$$K_w = [H^+] [OH^-]$$
$$= (10^{-7})(10^{-7})$$

The constant, K_w, is called the **ion product constant of water,** and on the p-scale

$$pK_w = pH + pOH$$
$$= 14$$

This last relationship will be useful for calculating the pH of basic solutions once the $[OH^-]$ is determined.

Acid Strength and Acid Concentration

Students must clearly understand the difference between acid *strength* and *concentration.* Strong acids and bases dissociate completely, weak ones only partially. HCl is a strong acid, whether it is concentrated or dilute because in its reaction with water, the equilibrium lies well to the right.

$$HCl + H_2O \longrightarrow H_3O^+ + Cl^-$$

Acetic acid is a weak acid whether it is concentrated or dilute because in its reaction with water, the equilibrium lies well to the left.

$$CH_3COOH + H_2O \rightleftharpoons CH_3COO^- + H_3O^+$$

Concentration of solutions, including acids and bases, was discussed in Chapter 2. For example, 6.0 mol/L HCl is more concentrated than 0.2 mol/L HCl. Also, 0.2 mol/L solutions of HCl and acetic acid are of the same *concentration;* however, the former adds a lot more H^+ ions to water than the latter.

The generalised weak acid HA ionises as follows:

$$HA \rightleftharpoons H^+ + A^-$$

The dissociation or equilibrium constant, K_{eq}, for acid ionisations is also called K_a (**acidity constant**):

$$K_a = \frac{[H^+][A^-]}{[HA]}$$

As greater dissociations produce more $[H^+]$, the acidity constant is a measure of the strength of the acid. pK_a is the corresponding $-\log$ term. Pyruvic acid with a K_a of 4.07×10^{-3} mol/L (pK_a 2.39) is a stronger acid than acetic acid, whose K_a is 1.74×10^{-5} mol/L (pK_a 4.6). Note that with molecules that can donate more than one proton, there will be a corresponding number of K_a and pK values. This is a feature of the amino acids, where the α-amino, the α-carboxy and, when present, the R-group, can each have its own pK.

Buffers

A buffer solution is one that resists a change in pH on the addition of an acid or alkali. Buffers are solutions of a weak acid and one of its salts (the conjugate base) or a weak base and one of its salts (a conjugate acid). The most useful equation for dealing with quantitative aspects of buffers is the **Henderson-Hasselbalch equation:**

$$pH = pK_a + \log_{10} \frac{[A^-]}{[HA]}$$

All the terms in this equation have been previously described. The [ ] strictly mean molar concentrations, but as a ratio is involved, any *chemical* concentration unit will do. Students must gain a good feel for the whole equation as well as the individual terms; it can be quite tricky because ratios and negative logs are involved and also because chemical terms, such as *base* and *salt,* could mean the same thing. Please check the following:

- pH is the variable term, the pH of the required buffer.
- pK_a is fixed once the buffer system is chosen (e.g., for acetate buffer, $pK_a = 4.76$).
- $\dfrac{[A^-]}{[HA]} = \dfrac{[salt]}{[acid]} = \dfrac{[base]}{[acid]} = \dfrac{[non\text{-}protonated]}{[protonated]}$

  The last expression is particularly descriptive when several protonated or deprotonated groups are involved, such as with amino acid structures.
  The ratio $\frac{[A^-]}{[HA]}$ or $\frac{[salt]}{[acid]}$ is the most important term in the Henderson-Hasselbalch equation. It is this term that determines the pH of the buffer. It is this term that students should carefully evaluate in buffer calculations and preparations. For a start, note that log of $1 = 0$, log of a number > 1 is positive, and log of a number < 1 is negative. It therefore follows that:

  - when [salt] = [acid], pH = pK (with the acetate buffer system, the pH of the buffer would be 4.76)
  - when [acid] > [salt], pH < pK (acetate buffer, pH would be < 4.76)
  - when [acid] < [salt], pH > pK (acetate buffer, pH would be > 4.76).

  The simple rule of thumb is: *the greater the [acid], the lower the pH of the buffer.* Remember, lower pH means more acid.

- For effective buffering, there is a limit over which the ratio $\frac{[A^-]}{[HA]}$ can be varied; it can only be 10:1 either way, giving a log value ±1. That is, buffers only work ±1 unit on either side of their pK values.
- Besides their names (acetate, phosphate, etc.), *two* further parameters are required in describing buffers: (i) pH and (ii) concentration. Thus, we can have:
  - pH 4.6 2.0 mol/L acetate buffer
  - pH 4.6 0.2 mol/L acetate buffer
  - pH 7.0 0.5 mol/L phosphate buffer
  - pH 7.2 0.5 mol/L phosphate buffer.

Table 5.2   pK$_a$ Values of Commonly Used Buffers in Biochemistry

Acid or Base	pK
Acetic acid	4.75
Carbonic acid	6.10, 10.22
Citric acid	3.10, 4.76, 5.40
Glycine	2.35, 9.78
Phosphoric acid	1.96, 6.70, 12.30
HEPES = N-2-hydroxyethylpiperazine- N'-2-ethanesulfonic acid	7.50
PIPES = piperazine-N-N'-bis (2-ethanesulfonic acid)	6.80
TRIS = 2-amino-2-hydroxymethylpropane-1,3-diol	8.14

Note: The names that are hard to pronounce came about as result of a search for good biological buffers that are effective around pH 7. For an excellent discussion on criteria of buffers suitable for biological research, the student should read Wilson and Goulding, *A Biologist's Guide to Principles and Techniques of Practical Biochemistry;* for recipes on the preparation of buffers, refer to Dawson et al., *Data for Biochemical Research.*

- Buffers resist pH changes; this is called **buffer action.** The magnitude of the buffer action is called the **buffer capacity (β).** It is measured by the amount of strong base required to alter the pH by one unit:

$$\beta = \frac{\delta[\text{base}]}{\delta[\text{pH}]}$$

where $\delta[\text{pH}]$ is the increase in pH resulting from the addition of $\delta[\text{base}]$.

The buffer capacity is best at pH = pK; also buffer capacity naturally depends on the concentration of the buffer, as the concentration would determine the amount of species present — to be protonated or deprotonated. As noted earlier, some compounds have more than one proton donating or accepting groups, hence have more than one pK value and buffering range. Glycine is an example; the buffering ranges being pH 2.35 ± 1 and 9.78 ± 1. Some of the commonly used buffers in biochemistry are listed in Table 5.2.

The pH-Titration Curve

An understanding of the pH-titration curve is imperative to all buffer work. Some very important concepts, using the glycine pH-titration curve (Figure 5.1) as an illustration, are set out below:

- The titration curve was obtained by recording the pH while two aliquots of 10 mL 0.2 mol/L glycine were titrated, the first against 0.1 mol/L NaOH and the second against 0.1 mol/L HCl. A composite graph of the two titrations is shown.

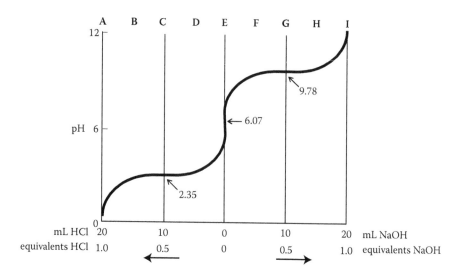

A all $^+NH_3CH_2COOH$	F $[^+NH_3CH_2COOH^-] > [NH_2CH_2COO^-]$
B $[^+NH_3CH_2COOH] > [^+NH_3CH_2COO^-]$	G $[^+NH_3CH_2COO^-] = [NH_2CH_2COO^-]$
C $[^+NH_3CH_2COOH] = [^+NH_3CH_2COO^-]$	H $[^+NH_3CH_2COO^-] < [NH_2CH_2COO^-]$
D $[^+NH_3CH_2COOH] < [^+NH_3CH_2COO^-]$	I all $NH_2CH_2COO^-$
E all $^+NH_3CH_2COO^-$	

Note: The uncharged form, NH_2CH_2COOH, does not appear at any pH value.

Figure 5.1   The pH-titration curve of glycine and the relative distribution of the different ionic species.

- Glycine goes into solution around its pI value, 6.07. The pH of glycine is extremely unsteady around this pH and can be markedly affected by the pH of water and slightest contaminations of the glassware.
- At the start, the glycine is in the zwitterion form, $^+NH_3CH_2COO^-$. Do not call this the *fully neutral* structure or anything implying *full-neutrality,* as these terms are used to refer to the titration curve and not to the nature of the ionic species.
- As NaOH is added, the equation below moves progressively to the right.

$$^+NH_3CH_2COO^- + NaOH \longrightarrow NH_2CH_2COO^-Na^+ + H_2O$$

H^+ is removed from $-NH_3^+$, forming water. Compare this reaction with:

$$HCl + NaOH \longrightarrow NaCl + H_2O$$

Students will recall referring to acid-base titrations in which salt and water are formed as **neutralisation.**

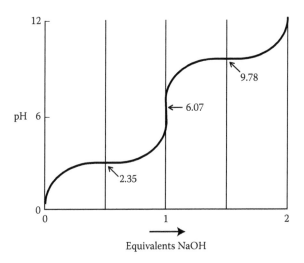

Figure 5.2   pH-titration of "diprotic" glycine.

- When one (whole) *equivalent* of NaOH is added, all the zwitterion $^+NH_3CH_2COO^-$ is converted into the anion $NH_2CH_2COO^-$. The word *equivalent* here means an equal amount of NaOH required to *fully neutralise* all of the $^+NH_3CH_2COO^-$ to salt and water. One *equivalent* here is 20 mL 0.1 mol/L NaOH; it is equivalent to 10 mL 0.2 mol/L glycine.
- The term *full-neutrality* is used to refer to the situation when the one equivalent (i.e., 20 mL NaOH) has been added (and the salt and water formed).
- Correspondingly, *half-neutrality* is when half-equivalent (or 10 mL NaOH) is added.
- At full-neutrality, all $^+NH_3CH_2COO^-$ is in the form $NH_2CH_2COO^-$, and at half-neutrality, only half is in the form $NH_2CH_2COO^-$ (the remainder is in the form $^+NH_3CH_2COO^-$).
- Remember, under conditions when an acid (here, $^+NH_3CH_2COO^-$) and its conjugate base (here, $NH_2CH_2COO^-$) are equal:

$$pH = pK + \log \frac{[base]}{[acid]}$$

the log term is zero, and pH = pK. That is, the pK is the pH value at exactly this situation.
- Unfortunately, this situation is also referred to as *equivalence point*; the reference is to the fact that $[^+NH_3CH_2COO^-]$ and $[NH_2CH_2COO^-]$ are equal. Do not confuse this equivalence point with the equivalent amount of NaOH needed for full-titration (i.e., the amount of NaOH needed to completely neutralise the acid, $^+NH_3CH_2COO^-$).
- To obtain the pK value for the $-NH_3^+$ of glycine, it is mathematically accurate to simply read the pH off the titration curve exactly at the point when 10 mL (or half-equivalent) NaOH has been added. Some textbooks assume this point to be the *point of inflexion* and assert that the pK occurs at the point of inflexion. It is difficult to determine the point of inflexion on a graph, and it may or may not correspond to the half-neutrality point on the curve. There could be practical errors elsewhere on the titration curve, resulting in

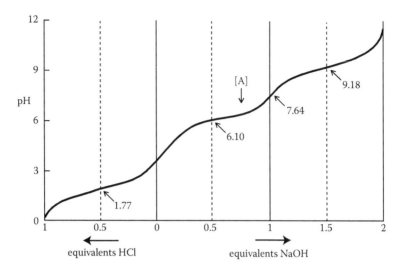

Figure 5.3   pH-titration curve of histidine. (Position A refers to Question 5.13.)

a lack of symmetry of the up-turn and down-turn parts of the curve, which in turn will make the point of inflexion unreliable. Furthermore, particularly when more than two pK values are present, as with certain amino acids and peptides, the symmetry of curves almost always fails (see the histidine titration curve, Figure 5.3). Curves are only pictorial representations. In this case, it is clear that taking the mathematical approach (i.e., reading off the pH corresponding to the 10 mL NaOH addition) would provide a more accurate estimation of the pK.

- It is at pH values close to the pK (i.e., pK ± 1) that the best buffer range is evident.
- The above comments also apply to the HCl titration, except this time the titration is:

$$^+NH_3CH_2COO^- + HCl \longrightarrow NH_3(Cl)CH_2COOH$$

In this case, as a neutral salt, $NH_3(Cl)CH_2COOH$, is formed, the term *neutralization* is again appropriate.

- Glycine, we know, contains an acid group (–COOH) and a basic group (–NH$_2$); but in the zwitterion structure, it is the –NH$_3^+$ group that has protons to donate and titrates with NaOH, while the –COO$^-$ accepts protons from the HCl. (The –COOH is not present and therefore cannot donate protons. This can be proven in an old organic chemistry experiment: titrating glycine in the presence of formaldehyde.)
- Indeed, two separate curves are obtained when free glycine is titrated against base and acid. As explained, this is due to the zwitterion nature of glycine. In Figure 5.1, the two curves are shown adjacent to each other. Many textbooks take the easy way out and consider a fully protonated cation or "diprotic" form of glycine, $^+NH_3CH_2COOH$, and show the progressive and eventually complete removal of two equivalents of protons by two equivalents of base, as shown in Figure 5.2. This approach is correct theoretically and is a convenient way of showing a third and, if necessary, further protons being removed (see Figure 5.5), by additional equivalents of NaOH.

pH-Titration Curve of Histidine

- Glycine contains only two ionisable groups: the α-amino and α-carboxy. Other amino acids may contain a third group — the R-group — also capable of ionisation. (The term *side chain* is best restricted for use in the polypeptide context.) The pH-titration of histidine, an example of an amino acid containing three ionisable groups, is shown in Figure 5.3.
  The R group of histidine, the imidazole group, has a pK in the acid range (see Table 5.3).
- Two equivalents of NaOH are needed to deprotonate the imidazole and $\alpha -NH_3^+$ groups, and one equivalent of HCl is needed to protonate the $\alpha -COO^-$ group. The resulting ionic structures, together with pK values, are shown in Figure 5.4.
- Alternatively, as some textbooks may show, three equivalents of NaOH are required to deprotonate a fully protonated, or "triprotic," histidine (as shown in Figure 5.5).

[Carefully look through the pK values in Table 5.3, not only for histidine but also for the other amino acids. Try to figure out at what pH values the groups will be protonated or deprotonated and in what order and how the groups would respond to NaOH/HCl titration.]

Table 5.3   pK and pI Values of Amino Acids

	pK α-COOH	pK α-NH$_3$	pK R-group	pI
Alanine	2.34	9.87		6.10
Glycine	2.35	9.78		6.07
Isoleucine	2.32	9.76		6.04
Leucine	2.36	9.60		6.03
Methionine	2.28	9.21		5.74
Phenylalanine	2.58	9.24		5.91
Proline	1.99	10.60		6.30
Serine	2.21	9.15		5.68
Threonine	2.15	9.12		5.63
Tryptophan	2.38	9.39		5.88
Valine	2.32	9.62		6.00
Aspartic Acid	2.09	9.82	3.87	2.96
Glutamic Acid	2.19	9.67	4.25	3.08
Histidine	1.77	9.18	6.10	7.64
Cysteine	1.71	10.78	8.33	5.02
Tyrosine	2.20	9.11	10.07	5.63
Lysine	2.18	8.90	10.58	9.71
Arginine	2.02	9.04	12.48	10.77

Note: These values are 'best approximations'. Values from different sources vary considerably. The pK values which contribute towards calculating the pI are in bold.

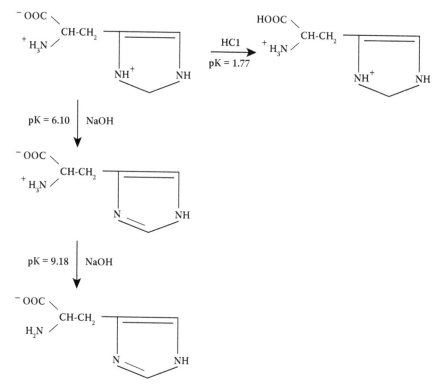

Figure 5.4   Ionic structures of histidine.

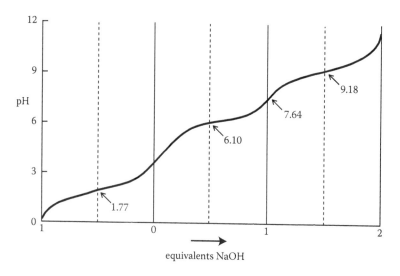

Figure 5.5   pH-titration curve of "triprotic" histidine.

Preparation of Buffers

In the preparation of buffers, two specifications must be met:

(a) the concentration of the buffer
(b) the pH of the buffer.

Remember, the concentration is the total material (acetate, phosphate, etc.) and the pH is obtained by having part of the total material as the protonated species (acid) and the remainder nonprotonated (salt or base). The base or salt can be either added (as, for example, Na acetate is added to acetic acid) or created (e.g., when NaOH is added to acetic acid, part of the acetic acid will become nonprotonated generating the salt Na^+Ac^-).

Bearing this in mind, there are three approaches to the preparation of buffers:

1. full-calculation method
2. calculation-titration method
3. data book method

In *the full-calculation method,* the amounts of acid and base are calculated (including using the Henderson-Hasselbalch equation) and mixed. The base may be added (see Question 5.10) or created (see Question 5.11).

The calculation-titration method, advocated by Dennison (1988), requires that only the total buffer material be calculated and the deprotonated species is simply generated during pH-titration with concentrated strong base (e.g., NaOH); the addition of the strong base is stopped when the required pH is reached and then diluted with water to give the final concentration of the buffer. While the computation part of the Henderson–Hasselbalch equation is not needed, its principles are nevertheless involved. Students must try to visualise the pH-titration curve, the progression as NaOH is added, the half-neutralization, the buffering region, the nature of the ionic species, and so forth (see Question 5.16).

Finally, in *the data book method,* tables of values calculated using the Henderson-Hasselbalch equation are provided in data books (e.g., Dawson et al., 1990). These tables provide a list of ratios of acid and base (either as salt or salt to be created) for a range of pH values (of the buffer). It is simply a matter of mixing the two parts together at the appropriate concentration (see Questions 3.21 and 3.22).

It is important when following the full-calculation or data book method that the pH of the buffer be checked on a pH meter (or at least with narrow-range pH paper). What if the pH is not exactly the expected value? Minor adjustments can readily be carried out on the pH meter by adding concentrated acid or base (salt) drop-wise while stirring.

Calculating pI of Amino Acids (and Peptides)

- Organic ionisable groups may become protonated by having protons (positive entities) incorporated into their structures. It is always only one proton that is either added or removed from a group. Hence protonation can render a group +1 (if it was 0 before) or 0 (if it was −1 before).
- pI is (the pH) when the molecule as a whole is neutral — that is, when the positive charges and negative charges balance each other. With proteins, this occurrence may involve several groups, but with amino acids it is always one positive against one negative. (See ionic structures of glycine, Figure 5.1.)

- With simple amino acids (those having only two ionisable groups), we determine the pI by finding the midpoint between the two pK values; we average the two pK values (see Table 5.3). What we are actually doing is averaging the pK, which makes the molecule go from +1 to 0 or from 0 to −1 (please look at Figure 5.1 again) — in other words, the pK values on either side of the "neutral," or net charge 0, molecule.
- With more complex amino acids (those having a third ionisable group), the difficult part is determining which of the pKs fit these specifications.
- Remember, we want the pK that makes the molecule overall +1 going to 0 and the pK that makes the molecule overall 0 going to −1. The easiest way to see how and when this is occurring is to visualise the pH titration of the fully protonated amino acid having its protons progressively stripped off as we add NaOH (like in Figures 5.2 and 5.5).
- With only three ionisable groups, the fully protonated species will have to be +2 (with basic amino acids) and +1 (with acidic amino acids). So when do they become 0? With basic amino acids, it will have to be +2 → +1 → 0 (i.e., at the second pK). With acidic amino acids, when it goes +1 → 0 (i.e., at the first pK). What about 0 to −1? We can gain quite a clear understanding by estimating the net charges on the molecule at selected pH values. (Of particular interest is when the net charge is zero.)
- Let us try to formulate some rules (for calculating pI values):

  - For amino acids with only two ionisable groups, simply average their pK values.
  - When there are three ionisable groups, the middle pK (in terms of numerical value) is always one of the pK values to be used in our averaging formula.
  - What about the other one? This one is a bit tricky. With acidic amino acids, the other one is the lowest (value) pK, and with basic amino acids, it is the highest (value) pK (see Table 5.3). It is not easy to generalise for the other amino acids.
  - The "all-embracing" rule is: determine the pH when the molecule is supposed to have a net charge of 0 (zero) and then average the pK values on either side of this pH.

- Deciding what charge a group bears at different pH values is easy. At a pH below the pK, the group in question is protonated (and deprotonated at pH values above the pK). Remember, pH is equal to pK at half-neutrality. First decide if the group is protonated; then decide whether it is 0 or +1. (For an illustration of the above features to selected amino acids, please see Question 5.14.)
- The above discussion has centred around amino acids. In peptides, there is an ionisable α-amino group at one end and an ionisable α-carboxy group at the other. Some of the R (or *side chain*) groups could be ionisable as well. Characteristic pH-titration curves, buffering capacities, pK, pI, and net charges are also a feature of peptides. (Application of these features to selected peptides is the basis of Question 5.15.)

Questions

1. Calculate pH when $[H+] = 5 \times 10^{-4}$.

2. Calculate pH of 0.02 mol/L HCl.

3. What is the pH of a 0.01 mmol/L HCl?

4. Calculate pH of 0.02 mol/L NaOH.

5. Calculate pH when 25 mL 0.16 mol/L NaOH is mixed with 50 mL 0.1 mol/L HCl.

6. How would you prepare a pH 3.3 HCl solution from concentrated (11.6 mol/L) HCl?

7. How would you prepare a pH 9.4 NaOH solution? M_r NaOH 40

8. How would you prepare a 0.1 mol/L pH 9.4 NaOH solution? M_r NaOH 40

9. What is the pH of a solution that is 0.4 mol/L Na lactate and 0.2 mol/L lactic acid? (pK_a lactic acid = 3.73)

10. What is the pH when the following solutions of acetic acid (pKa 4.76) are mixed?
    (a) 5 mL 0.1 mol/L acid + 5 mL 0.1 mol/L Na acetate
    (b) 3 mL 0.1 mol/L acid + 7 mL 0.1 mol/L Na acetate
    (c) 7 mL 0.1 mol/L acid + 3 mL 0.1 mol/L Na acetate
    (d) 6 mL 0.1 mol/L acid + 7 mL 0.1 mol/L Na acetate
    (e) 3 mL 0.2 mol/L acid + 7 mL 0.1 mol/L Na acetate

11. 100 mL of a 0.4 mol/L acetic acid is mixed with 20 mL of a 1.8 mol/L NaOH. Calculate the pH of this buffer. (pKa acetic acid = 4.76)

12. Describe the full preparation of 1.0 L 0.2 mol/L pH 4.4 acetate buffer. (pK_a = 4.76, M_r Na acetate 82, glacial acetic acid is 17.4 mol/L)

13. With the aid of Table 5.3, choose an amino acid that would provide a suitable buffer at pH 7.0 when made up with NaOH. How many grams of this amino acid and what volume of 0.1 mol/L NaOH would you mix to make 500 mL of a 0.01 mol/L solution of this buffer? (*Hint:* Having mixed the amino acid and the NaOH, you are expected to dilute to a final volume of 500 mL with distilled water.)

14. For each of the amino acids listed below, perform tasks (a) to (d) that follow:
    • valine
    • aspartic acid
    • arginine
      (a) Draw the structure and assign the pK values from Table 5.3 to each of the ionisable groups.
      (b) Tabulate the charges each ionisable group will bear at pH 2, 5, 7, 10, and 13.
      (c) What would be the net charge of each amino acid at pH 2, 5, 7, 10, and 13?
      (d) Estimate the pI for each amino acid.

15. Assume that the pK values for ionisable groups when they are part of a peptide are the same as when they are part of free amino acids. (The pK values do deviate from those of the free amino acid, but the reasons are complex and varied and will not be considered here.) For each of the peptides listed below, perform tasks (a) to (d) that follow:
    • ala-lys
    • ser-asp-glu
    • cys-glu-lys-arg
      (a) Draw the structure of the peptide and assign the pK values from Table 5.3 to each ionisable group.
      (b) Tabulate the charges each ionisable group will bear at pH 2, 5, 7, 10, and 13.
      (c) What would be the net charge of each peptide at pH 2, 5, 7, 10, and 13?
      (d) Estimate the pI for each peptide.

16. Ion exchange ionic-strength gradient elution is used in eluding ion exchange columns at a fixed pH with increasing amounts (gradient) of a salt (usually NaCl). In the procedure, the two chambers of a gradient generator are filled with different solutions and allowed to be reciprocally mixed. One chamber contains a buffer with a low concentration or no NaCl (the starting buffer), while the other contains a higher concentration of NaCl (the finishing buffer).

Using the calculation-titration approach discussed earlier, describe the preparation of 1 L of an ionic-strength gradient finishing buffer, pH 7.6, 0.1 mol/L Na-phosphate buffer, containing 1.0 mol/L NaCl.

$$M_r \; NaH_2PO_4 \cdot 2H_2O \; 156 \qquad NaCl \; 58.5$$

Answers

1. $pH = \log [H+] = -\log (5 \times 10^{-4}) = \mathbf{3.30}$

Students should be able to enter numbers as scientific notation into their electronic calculators and obtain negative $\log_{10}$ directly. To gauge if your answer is the right order of magnitude, please read the comments following answer 5. [On your electronic calculators, you are aware of course that 5×10^{-4} (which equals 0.0005) comes out as (5e −4).]

2. $0.02 \; mol/L = 2 \times 10^{-2} \; mol/L$

$$pH = -\log (2 \times 10^{-2})$$
$$= \mathbf{1.70}$$

3. Molar concentration $= 0.01$ mmol/L

$$= 0.000 \; 01 \; mol/L$$
$$= 1 \times 10^{-5} \; mol/L$$
$$pH = \mathbf{5}$$

4. $[OH^-] = 0.02 = 2 \times 10^{-2}$

$$pOH = -\log (2 \times 10^{-2}) = 1.70$$
$$pH = 14 - 1.70$$
$$= \mathbf{12.30}$$

5. mmol NaOH $= 25 \times 0.16 = 4$

mmol HCl $= 50 \times 0.1 = 5$

$$NaOH + HCl \longrightarrow NaCl + H_2O$$

4 mmol 5 mmol   4 mmol 4 mmol

1 mmol HCl is excess in a total volume of 75 mL

$$= 1 \text{ mmol} / 75 \text{ mL}$$

$$= 0.0133 \text{ mmol/mL}$$

$$= 0.0133 \text{ mol/L}$$

$$pH = -\log 0.0133$$

$$= \mathbf{1.88}$$

Comments: Let's have a quick check of our understanding of $-\log$, pH, and $[H^+]$.
If 0.01 mol/L HCl, pH = 2.
If 0.1 mol/L HCl, pH = 1.
As 0.0133 mol/L is between 0.01 and 0.1, then pH should be between 1 and 2 (i.e., 1.88 is correct).

6. First calculate the $[H^+]$ when pH = 3.3

$$pH = -\log [H^+]$$

$$3.3 = \log 10^{-3.3}$$

$$\text{antilog of } 10^{-3.3} = 5 \times 10^{-4}$$

$$[H^+] = 5 \times 10^{-4} \text{ (can be done with fewer steps on an electronic calculator)}$$

We need 5×10^{-4} mol/L HCl.
We need to dilute the 11.6 mol/L HCl to give 5×10^{-4} mol/L.
The extent of dilution would be:

$$\frac{11.6}{5 \times 10^{-4}} = 23\,200$$

We need a 1 : 23 200, dilution which is too much to carry out in one step. A two-step dilution, as discussed in Chapter 3, is required.

$$1.0 \text{ mL } 11.6 \text{ mol/L diluted to } 116 \text{ mL} = 0.1 \text{ mol/L}$$

$$\text{then } 1.0 \text{ mL} \rightarrow 200 \text{ mL would give us } 200 \text{ mL HCl of pH } 3.3.$$

Note: This dilution can be checked quite easily:

$$\frac{1}{116} \times \frac{1}{200} = \frac{1}{23200}$$

which is the dilution required.

7. pOH = 14 − 9.4

$$= 4.60$$

$$[OH^-] = 10^{-4.6}$$

$$\text{antilog } 10^{-4.6} = 2.5 \times 10^{-5}$$

$$\text{Need } 2.5 \times 10^{-5} \text{ mol/L NaOH}$$

$$1 \text{ mol/L} = 40 \text{ g/L}$$

$$2.5 \times 10^{-5} \text{ mol/L} = \frac{2.5 \times 10^{-5}}{1} \times 40$$

$$= \mathbf{0.001 \ g/L}$$

8. It is not possible to prepare this solution as the concentration would dictate the pH and vice versa. When pH is 9.4, [NaOH] is 2.5×10^{-5} (Question 7). When [NaOH] is 0.1 mol/L,

$$\text{pOH} = -\log 10^{-1}$$

$$= 1$$

$$\text{pH} = 14 - 1$$

$$= \mathbf{13}$$

Note: Students should know just by inspection that pH of 0.1mol/L HCl is 1, and (therefore) pH of 0.1 mol/L NaOH is 13.

9. $\text{pH} = \text{pKa} + \log \dfrac{[\text{salt}]}{[\text{acid}]}$

   $= 3.73 + \log \dfrac{0.4 \text{ mol} / L}{0.2 \text{ mol} / L}$

   $= 3.73 + \log 2$

   $= 3.73 + 0.30$

   $= \mathbf{4.03}$

10. (a) $\text{pH} = \text{pK} = \mathbf{4.76}$; the log term will be zero because salt and acid are equal.

    (b) $\text{pH} = \text{pK} + \log \dfrac{[\text{salt}]}{[\text{acid}]}$

    $= 4.76 + \log \dfrac{7}{3}$

    $= 4.76 + \log 2.33$

    $= 4.76 + 0.37$

    $= \mathbf{5.13}$

(c) $pH = pK + \log \dfrac{[salt]}{[acid]}$

$$pH = pK + \log \dfrac{3}{7}$$

$$= 4.76 + \log 0.43$$

$$= 4.76 - 0.37$$

$$= \mathbf{4.39}$$

Note: When acid > salt, pH is lower than pK.

(d) $pH = pK + \log \dfrac{7}{6}$

$$= 4.76 + \log 1.17$$

$$= 4.76 + 0.07$$

$$= \mathbf{4.83}$$

(e) $pH = pK + \log \dfrac{7 \times 0.1}{3 \times 0.2}$

$$= \mathbf{4.83} \text{ [same answer as (d)]}$$

Note: Please inspect the answers to all five parts of the question. In high-school formula approach, you would be expected to work out the moles of acid and salt in each situation, obtaining figures such as 0.0003 and 0.0007 or even waste time with 3×10^{-4} and 7×10^{-4}. At the more mature level, simply write $\frac{3}{7}$ in ratios (and be done with it!). The same maturity and authority over mathematical and chemical principles will make you realize that when materials go into the same final volume, the final volume does not have to feature in the calculations — the ratio $\frac{6}{13} : \frac{7}{13}$ is simply 6 : 7; and for the same volume, there are twice as many moles when the concentration is 0.2 mol/L instead of 0.1 mol/L.

11. This situation is like a titration of HAc by NaOH:

$$HAc + NaOH \rightleftharpoons NaAc + H_2O$$

A portion of the HAc will be converted to NaAc. This salt will then remain as a mixture with the remainder of the HAc. Remember the equilibrium of HAc is very much to the left, so we can ignore any acetate ions at the start.

$$HAc \rightleftharpoons Ac^- + H^+$$

Na$^+$ ions will *assist* the equilibrium to the right and *capture* the Ac$^-$ ions formed as NaAc.

$$\text{Amount of NaAc formed} = \text{Amount of NaOH added}$$

$$= 1.8 \times 20 \text{ mmol}$$

$$= 36 \text{ mmol}$$

$$\text{Amount of HAc at start} = 0.4 \times 100 \text{ mmol}$$

$$= 40 \text{ mmol}$$

$$\text{Amount of free HAc remaining} = 40 - 36 = 4 \text{ mmol}$$

$$\text{Ratio } \frac{\text{Ac}^-}{\text{HAc}} = \frac{36}{4} = 9$$

$$\text{pH} = \text{pK} + \log 9$$

$$= 4.76 + 0.95$$

$$= \mathbf{5.71}$$

12. Solutions of 0.2 mol/L acetic acid and 0.2 mol/L sodium acetate have to be prepared and mixed at a particular ratio so that the final volume is 1.0 L and the pH 4.4. The concentration will be 0.2 mol/L if 0.2 mol/L and 0.2 mol/L solutions are mixed.

$$\text{pH} = \text{pK} + \log \frac{[\text{salt}]}{[\text{acid}]}$$

$$4.4 = 4.76 + \log \frac{[\text{salt}]}{[\text{acid}]}$$

$$(= -0.36)$$

i.e., $\log \dfrac{[\text{salt}]}{[\text{acid}]} = -0.36$

It is easier to deal with log = 0.36.

$$\text{anti log} = 2.291$$

i.e., $\dfrac{[\text{acid}]}{[\text{salt}]} = \dfrac{2.291}{1}$

[The approach with electronic calculators may be different, but we must get to this expression or the inverse of this expression.]

The volumes of acid and salt must add up to 1000 mL.

Let x be the number of mL of salt:

$$2.291x + 1x = 1000 \text{ mL}$$

$$3.291x = 1000 \text{ mL}$$

$$x = 304.1 \text{ mL}$$

Volumes required: **304 mL 0.2 mol/L Na acetate + 696 mL 0.2 mol/L acetic acid.**

Note: The volume of acid is greater than the volume of salt as we require a pH lower than the pK.

Do not forget, *you* have to make these solutions:

M_r Na acetate 82

$$8.2 \text{ g}/500 \text{ mL} = 0.2 \text{ mol/L}$$

Glacial acetic acid is 17.4 mol/L; we want 0.2 mol/L.

$$\text{Ratio is } \frac{17.4}{0.2} = 1 : 87$$

$$\Rightarrow 10 \text{ mL} \rightarrow 870 \text{ mL} = 0.2 \text{ mol/L}$$

Now mix 304 mL and 696 mL (respectively) of the two solutions, and check the pH of the prepared buffer on a pH meter.

13. We are looking for the closest pK to 7.0 (not the closest pI; there is minimum pH resistance at around the pI — have a look at Figure 5.1 and 5.3). The two candidates are cysteine (pK of R-group = 8.33) and histidine (pK of R-group = 6.10). Histidine is chosen because we want to get the pH *up* to 7.0, and adding NaOH (a base) is a sure way of increasing the pH! The R-group (imidazole) ionization of histidine is shown in Figure 5.6.

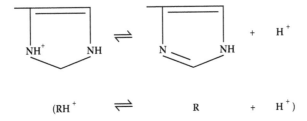

Figure 5.6   Answer to Question 5.13.

The imidazole group of histidine ionises as:

$$RH^+ + \rightleftharpoons R + H^+$$

With NaOH:

$$RH^+ + OH^- \rightleftharpoons R + H_2O$$

$$\uparrow \qquad\qquad \uparrow$$

(protonated) (nonprotonated)

The addition of NaOH will trap a certain amount of histidine as the nonprotonated species (R). See position [A] on the histidine titration curve (Figure 5.3).

[For obvious reasons, the M_r His = 155 was not given with this question; you had to find it elsewhere.]

To make 500 mL 0.01 mol/L His buffer, we require:

$$500 \times 0.01 = 5 \text{ mmol His}$$

$$= 5 \times 155$$

$$= 775 \text{ mg}$$

We want a portion of the 5 mmol His as RH^+ and remainder as R. The R is created by reacting His with NaOH:

$$pH = pK + \log \frac{[R]}{[RH^+]}$$

$$7.0 = 6.1 + \log \frac{[R]}{[RH^+]}$$

$$(= 0.9)$$

antilog $0.9 = 7.94$

i.e., $\dfrac{[R]}{[RH^+]} = \dfrac{7.94}{1}$

Of the 5 mmoles His, let x be the R form and $(5 - x)$ be the RH^+ form:

$$\frac{x}{5 - x} = \frac{[R]}{[RH^+]} = \frac{7.94}{1}$$

$$x = 7.94 \, (5 - x)$$

$$x = 39.7 - 7.94x$$

$$8.94x = 39.7$$

$$x = 4.441 \text{ mmol}$$

This is the R form and is created by using 4.441 mmol, or 44.41 mL, of a 0.1 mol/L solution of NaOH.

As we have said, the remainder $5 - 4.441 = 0.559$ mmol remains as RH^+.

Check:

$$\frac{4.441}{0.559} = \frac{7.94}{1}$$

Final solution (buffer) is made by **dissolving 775 mg His, adding 44.41 mL 0.1 mol/L NaOH, and diluting to 500 mL** — and don't forget to check the pH on the pH meter before using the buffer.

14. (a) The structures and pK values are shown in Figure 5. 7.
Answers to Parts (b), (c), and (d) are summarised in Table 5. 4, and detailed working and comments are set out below.
(c) Inspection of the net charges tells us the pI of Val should be between 5 and 7. Because there are only two pK values, you don't need to be Einstein to figure out which two pKs to average.

$$\textbf{Val pI} = \frac{2.32 + 9.62}{2} = \textbf{5.97}$$

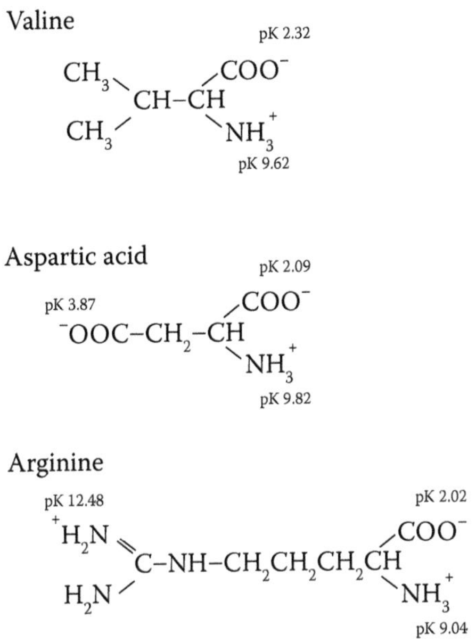

Valine

$$\underset{\substack{\text{pK 9.62}}}{\overset{\substack{\text{pK 2.32}}}{\underset{CH_3}{\overset{CH_3}{\diagdown}}CH-CH\overset{COO^-}{\underset{NH_3^+}{\diagup}}}}$$

Aspartic acid

$$\overset{\substack{\text{pK 3.87}}}{^-OOC-CH_2-}\underset{\substack{\text{pK 9.82}}}{CH\overset{\overset{\text{pK 2.09}}{COO^-}}{\underset{NH_3^+}{\diagup}}}$$

Arginine

$$\underset{H_2N}{\overset{\substack{\text{pK 12.48}}}{^+H_2N}}C-NH-CH_2CH_2CH_2\underset{\substack{\text{pK 9.04}}}{CH\overset{\overset{\text{pK 2.02}}{COO^-}}{\underset{NH_3^+}{\diagup}}}$$

Figure 5.7   Answer to Question 5.14a.

For Asp pI: Looking at Table 5.4, the 0 net charge is between pH 2 and pH 5. The pKs on either side are 2.09 and 3.87 (the *middle* pK). So, average these pK values:

$$\frac{2.09 + 3.87}{2} = 2.98$$

		Table 5.4	Answers to Question 5.14				
Amino Acid Group	Calculated pI	pK	pH 2	pH 5	pH 7	pH 10	pH 13
Valine	**5.97**						
α-COOH		2.32	0	–	–	–	–
α-NH₂		9.62	+	+	+	0	0
Net Charge			+1	0	0	–1	–1
Aspartic acid	**2.98**						
α-COOH		2.09	0	–	–	–	–
α-NH₂		9.82	+	+	+	0	0
β-COOH		3.87	0	–	–	–	–
Net Charge			+1	–1	–1	–2	–2
Arginine	**10.76**						
α-COOH		2.02	0	–	–	–	–
α-NH₂		9.04	+	+	+	0	0
Guanindino group		12.48	+	+	+	+	0
Net Charge			+2	+1	+1	0	–

For Arg pI: That's easy, it occurs around 10. So we average the pKs on either side: 9.04 (the *middle* one!) and 12.48:

$$\text{Average} = \frac{9.04 + 12.48}{2} = \textbf{10.76}$$

Note: The calculated pI values are included in Table 5.4.

15. (a) The convention when drawing structures of peptides is to start from the N-terminal. The structures and the pK values of ionisable groups are shown in Figure 5.8.

Answers to Parts (b), (c), and (d) are summarised in Table 5.5, and detailed working and comments are set out below.

Conveniently, the pK values of each ionisable group of each peptide is arranged in ascending order. It is easier to see patterns developing. Some pI values can be written down virtually by inspection. Follow the rules given in the introduction to this chapter.

For Ala-Lys:

$$pI = \frac{9.87 + 10.58}{2} = \textbf{10.23}$$

Table 5.5   Answers to Question 5.15

Peptide Ionisable Group	Calculated pI	pK	pH 2	pH 5	pH 7	pH 10	pH 13
Ala-Lys	**10.23**						
Lys α-COOH		2.20	0	−1	−1	−1	−1
Ala α-NH$_2$		9.87	+1	+1	+1	0	0
Lys R-group		10.58	+1	+1	+1	+1	0
Net Charge			**+2**	**+1**	**+1**	**0**	**0**
Ser-Asp-Glu	**2.92**						
Glu α-COOH		2.19	0	−1	−1	−1	−1
Asp R-group		3.65	0	−1	−1	−1	−1
Glu R-group		4.25	0	−1	−1	−1	−1
Ser α-NH$_2$		9.15	+1	+1	+1	0	0
Net Charge			**+1**	**−2**	**−2**	**−3**	**−3**
Cys-Glu-Lys-Arg	**9.46**						
Arg α-COOH		2.02	0	−1	−1	−1	−1
Glu R-group		4.25	0	−1	−1	−1	−1
Cys R-group		8.33	0	0	0	−1	−1
Cys α-NH$_2$		10.78	+1	+1	+1	+1	0
Lys R-group		10.58	+1	+1	+1	+1	0
Arg R-group		12.48	+1	+1	+1	+1	0
Net Charge			**+3**	**+1**	**+1**	**0**	**−3**

For **Cys-Glu-Lys-Arg:**

$$pI = \frac{8.33 + 10.58}{2} = \textbf{9.44}$$

Calculating the pI for **Ser-Asp-Glu** poses a problem. We want to average the pK values on either side of the species bearing a 0 net charge. ("0" does not actually appear in Table 5.5 — the net charge jumps from +1 to –2.) There are, unfortunately, three pK values in the range pH 2 to pH 5. We cannot decide whether to average the first and the second or the second and third. (No, we are *not* allowed to average all three!) There is too much of a jump from pH 2 to 5 for us to make a critical decision. Maybe, if we "zoom-in" on this range, we might get some clues. That's what is being done in Table 5.6.

	pK	pH 1	pH 2	pH 3	pH 4	pH 5
Glu α-COOH	2.19	0	0	–1	–1	–1
Asp R-group	3.65	0	0	0	–1	–1
Glu R-group	4.25	0	0	0	0	–1
Ser α-NH$_2$	9.15	+1	+1	+1	+1	+1
Net Charge		**+1**	**+1**	**0**	**–1**	**–2**

Table 5.6   Answer to Question 5.15 – Showing Expanded pH Range

From Table 5.6, we can clearly see that the pKs on either side of net charge 0 are **2.19** and **3.65,** and their average is **2.92** (which is the pI).

The pI values for all three peptides are entered in the main table (Table 5.5).

[Please take a close at the structures of the peptides in Figure 5.8. Consider the effects their amino acid compositions and "exposed" side chains have on the pI.]

16. Mass NaH$_2$PO$_4$.2H$_2$O required = 15.6 g for 1 L 0.1 mol/L
    Mass NaCl required = 58.5g for 1 L 1.0 mol/L
    In the calculation-titration approach, no complex calculations are required. However, note that phosphoric acid goes (see Table 5.2):

$$\overset{1.96}{} \qquad \overset{6.70}{} \qquad \overset{12.30}{}$$

$$\underset{NaOH}{H_3PO_4 \longrightarrow} \underset{NaOH}{NaH_2PO_4 \longrightarrow} \underset{NaOH}{Na_2HPO_4 \longrightarrow} Na_3PO_4$$

as the respective pK values are crossed in the titration against NaOH. The relevant transformation here is:

$$\overset{6.70}{}$$

$$\underset{NaOH}{NaH_2PO_4 \longrightarrow} Na_2HPO_4$$

A pH-monitored titration with NaOH can be stopped when the right ratio of $\frac{Na_2HPO_4}{NaH_2PO_4}$ is reached.

Ala-Lys

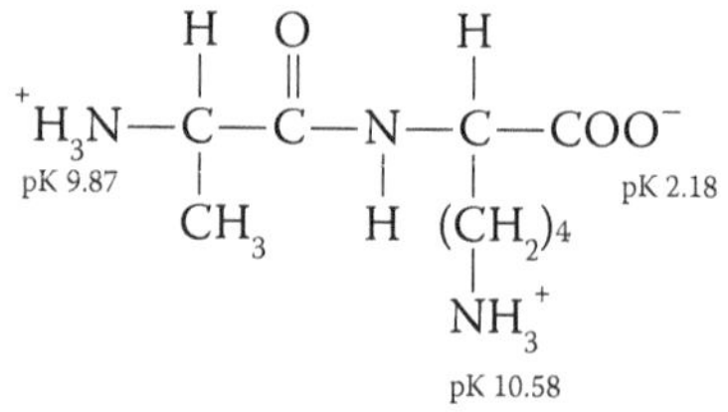

Ser-Asp-Glu

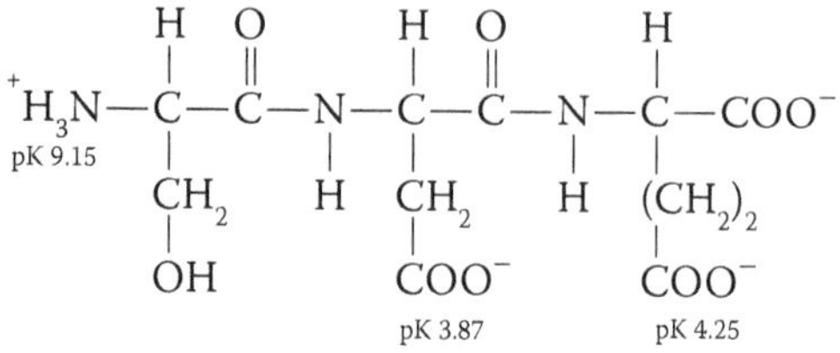

Cys-Glu-Lys-Arg

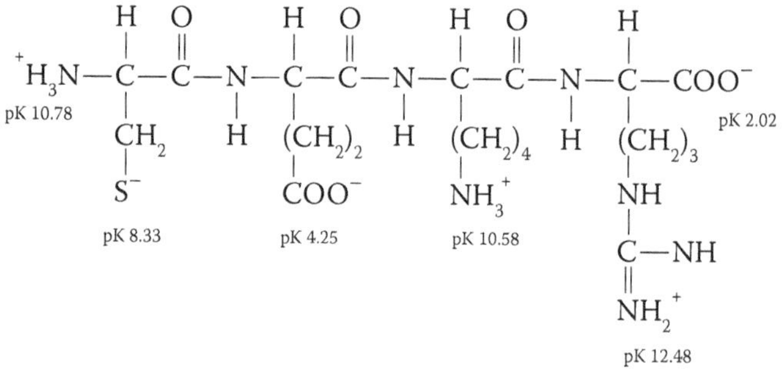

Figure 5.8   Answer to Question 5.15a.

Method: Dissolve the $NaH_2PO_4.2H_2O$ and the NaCl in about 900 mL distilled H_2O and titrate to pH 7.6 with a fairly concentrated NaOH (e.g., 10 mol/L or 40 % NaOH). Then make the final volume to 1 L with distilled H_2O.

6

Polarimetry

Polarimetry deals with the rotation of plane polarised light by optically active solutes in solution. As optically active pairs of substances have almost identical chemical properties, the technique of polarimetry afforded early organic chemists and biochemists a viable technique for differentiating, quantitating, and identifying this important group of compounds. Students are advised to refresh their knowledge of chiral carbon, stereoisomers, optical isomers, enantiomers (mirror images), diastereomers (glucose, galactose, mannose), epimers (glucose and galactose), anomers (α-D- and β-D-glucopyranose), and chair and boat conformations.

As enzymes are highly specific and able to distinguish between almost every pair of isomers, their use seems to have overshadowed the use of the polarimeters in most biochemical laboratories. Adding to the disappearance of polarimeters in modern biochemistry laboratories were the advances in spectrophotometers and mass spectrometers including, in both cases, their automation and control by computers. Is polarimetry, then, outdated? It was for several decades, but recently it is experiencing a renaissance as a useful tool in modern biotechnology. Part of its attraction is due to the nondestructive nature of its application. Any change in molecular structure produces different molecules (obviously) and each optically active compound has its unique rotation (Biot's Law). Hence, subtle changes currently investigated in modern molecular biology may be monitored with this technique. Among the applications, we have monitoring of antibody-antigen binding, nucleic acid hybridisation, detection of cancerous cells (the optical characteristics of compounds in living cells may actually be observed), and detection and analysis of the therapeutically active enantiomers in certain recent pharmaceutical compounds. It would also be of interest for students in food science to know that polarimeters (or *saccharimeters*) continue to play a central role in the sugar (sucrose) industry (see Question 6.7).

The quantitative work relating to polarimetry is treated in this chapter. For more detailed discussion on the basic aspects of this topic, students are referred to their organic chemistry textbooks.

The major formula dealing with polarimetric calculations is:

$$[\alpha]_D^T = \frac{\alpha}{lc} \tag{i}$$

where $[\alpha]_D^T$ is called the **specific rotation,** measured in degrees, of a 1 g of solute/mL in a 1 dm, or 10 cm, tube (The readings are taken as polarised light enters at one end and passes through the optically active solution in a 10 cm tube bounded by optically clear glass.) As the density of the solution and its optical activity vary with temperature, the temperature is denoted by superscript T but usually does not enter calculations. Neither does D (the D-line of sodium at 589 nm).

A specific wavelength is used as students will realise that longer wavelengths can be disturbed (rotated) more than shorter ones.

α is the observed rotation in degrees
l is the length of light path through the solution measured in decimetres
c is the concentration in g/mL of solution.

With this, as with any formula, students must obtain a feel for the overall expression as well as the individual variables. Equation (i) can be rearranged giving:

$$\alpha = \left[\alpha\right]_D^T lc \tag{ii}$$

This tells us (what is obvious) that the higher the concentration or the longer the tube, the greater the rotation. Note that the proportionality constant is the defined term $\left[\alpha\right]_D^T$.

An alternative form of the equation relating specific rotation is:

$$\left[\alpha\right]_D^T = \frac{100\alpha}{lc}$$

In this case, c is expressed in grams per 100 mL (g/100 mL) solution.

Questions

1. Molar rotation [M] is defined as the rotation of 1 mol/100 mL solution in a 1 dm tube. You are given the specific rotation of D-asparagine (M_r 132) as +5.42°:
   (a) Calculate [M]. **Note:** This is *not* the rotation of a 1 mol/L solution.
   (b) This time, calculate the rotation of a 1 molar (mol/L) solution in a 1 dm tube.

2. Calculate the amount of glucose in a 10 mL solution that gives an equilibrium rotation of +26.5° in a 2 dm tube. Refer to Table 6.1.

3. A 5% w/v solution of an unidentified sugar gave a rotation of +4.05° in a 1 dm tube. Which of the sugars given in Table 6.1 is this sugar likely to be?

Table 6.1   Specific Rotation of Sugars (Equilibrium Mixtures) in Water

L-Arabinose	+104.5°	D-Mannose	+14.6°
D-Xylose	+19.0°	L-Rhamnose	+8.9°
D-Galactose	+80.0°	Lactose	+55.3°
D-Glucose	+52.0°	Maltose	+136.0°
D-Fructose	−92.0°	Sucrose	+66.5°

Table 6.2 Specific Rotation of Amino Acids in Water

Alanine	+14.6°	Isoleucine	+39.5°
Arginine	+27.6°	Leucine	+16.0°
Aspartic acid	+25.4°	Proline	−60.4°
Glutamic acid	+31.8°	Serine	+15.1°
Glycine	0*	Tyrosine	+5.2°

* Why is glycine zero? Also, why is the specific rotation of succinic acid zero?

4. Calculate the rotation of a 0.5 mol/L sucrose (M_r 342) solution in a 1 dm tube. A few drops of concentrated HCl are added to the sucrose solution and heated to completely hydrolyse the sucrose. Assume there is no volume change. What sugars would be produced? What would be their concentration? What would be the rotation of this hydrolysate? Refer to Table 6.1.

5. A 30 g mixture of alanine and glutamic acid was dissolved in a final volume of 100 mL water. Their combined rotation in a 20 cm tube was determined to be +47°. Calculate the percentage of each amino acid. Refer to Table 6.2.

6. An equilibrium mixture of α- and β-D-glucose has a specific rotation of +52.7°. The specific rotations for α-D-glucose and β-D-glucose are +112° and +18°, respectively. Calculate the proportions of each anomer in the equilibrium mixture.

7. (This question may be of special interest to students in food science programs. Polarimeters (or *saccharimeters*) continue to play central role in the sugar (sucrose) industry. An International Sugar Scale (°Z) is used by the industry to accurately report the sugar purity. A "normal" standard 260 g/L gives a rotation of 34.626°; however, this is defined in the International Sugar Scale as 100°Z.)
   (a) How does the figure 34.626° compare with values given in Table 6.1?
   (b) A 24.632 g/100 mL sugar sample, taken from the final stages of sugar purification and concentration, gives a rotation of 32.633° (in a 20 mm tube at 20° and wavelength 589.44 nm — the same conditions under which the International Sugar Scale is defined. Assuming that the impurities themselves do not have any rotation, and they do not interfere with the sucrose rotation, calculate the % purity of the sample.

Answers

1. (a) 1 g/mL solution gives a rotation of +5.42°.

$$1 \text{ g/100 mL solution would give } \frac{5.42°}{100}.$$

$$\Rightarrow 132 \text{ g/100 mL would give } \frac{132}{1} \times \frac{5.42°}{100} \text{ *}$$

$$= \mathbf{7.15°}$$

* **Note** the formula for [M] is indeed: $[M] = \dfrac{M_r}{100}\alpha$

(b) 1 molar solution is 132 g/1000 mL or 13.2 g/100 mL. Hence, the α should be **0.715°**.

Check: In [M], the concentration unit is mol/100 mL, whereas in calculating the rotation of a 1 molar solution, the concentration unit is mol/1000 mL; hence, the latter is a more dilute solution.

2. **By formula method:**

$$c = \frac{\alpha}{[\alpha]_D^T 1} = \frac{26.5°}{52.08 \times 2}$$

$$= 0.255 \text{ g/mL}$$

$$= \textbf{2.55 g/10 mL}$$

By ratio method:

26.5° given in a 2 dm tube

$\dfrac{26.5°}{2}$ given in a 1 dm tube

52° given by a 1 g/mL solution

$\dfrac{26.5°}{2}$ given by a $\dfrac{26.5°}{2} \times \dfrac{1}{52°}$ g/mL solution

$$= 0.255 \text{ g/mL}$$

$$= \textbf{2.55 g/10 mL}$$

Note: The ratio method looks longer, but it will come in handy if one cannot remember the formula. Besides, not all the steps have to be written down. The ratios can be imagined and one final expression written down. Where there is a defined formula (such as in these polarimetry calculations) and the student can recall it, by all means, plug in the values and proceed. Be careful and evaluate the final answer.

3. 5% w/v = 5 g/100 mL = 0.05 g/mL

   0.05 g/mL gave a rotation of +4.1°

   $\Rightarrow$ 1 g/mL would give $\dfrac{1}{0.05} \times 4.05°$

   $$= \textbf{+81°}$$

   Answer: D-galactose

4. 1 g/mL sucrose, rotation is +66.5°

   1 g/L sucrose, rotation = $\dfrac{+66.5°}{1000}$

   0.5 mol/L = 171 g/L would give $\dfrac{+66.5°}{1000} \times 171$

   $$= \textbf{+11.4°}$$

Equimolar glucose and fructose would be produced; each would be 0.5 mol/L.

$$0.5 \text{ mol/L glucose} = 90 \text{ g/L, rotation} = \frac{+52}{1000} \times 90$$

$$= +4.68°$$

$$0.5 \text{ mol/L fructose} = 90 \text{ g/L, rotation} = \frac{-92.0}{1000} \times 90$$

$$= -8.28°$$

Final rotation $= (+4.68°) + (-8.28°) = -3.6°$

Notice that the rotation was positive before the hydrolysis and upon hydrolysis became negative (due to the strong negative rotation of fructose). In other words, the rotation was *inverted* to the other direction. Historically, this was the reason for the name *invert sugar,* which used today simply means an equimolar mixture of glucose and fructose. The yeast enzyme invertase owes its name to this phenomenon. The mammalian intestinal enzyme, however, is called *sucrase.*

5. Rotation in a 1 dm tube would be $+23.5°$ and 30 g/100 mL $= 0.3$ g/mL.

Let ala $= x$ g/mL

glu $= (0.3 - x)$ g/mL

In the final solution, each amino acid will contribute a rotation directly proportioned to its concentration.
Using specific rotations from Table 6.2:

$$x(+14.6) + (0.3 - x)(+31.8) = 0.3(23.5)$$

$$14.6x + 9.54 - 31.8x = 7.05$$

$$-17.2x = -2.49$$

$$x = 0.145 \text{ g}$$

$$\text{ala} = x = 0.145 \text{ g} = \textbf{48\%}$$

$$\text{gly} = 0.3 - x = 0.155 \text{ g} = \textbf{52\%}$$

6. The contribution of each anomer towards the final rotation is directly proportional to the concentration of each anomer in the equilibrium mixture.

Let $x = \% \ \alpha$

$\Rightarrow 100 - x = \% \ \beta$

$x(+112°) + (100 - x)(+18.7°) = 100(+52.7°)$

$112x + 1870 - 18.7x = 5270$

$93.3x = 3400$

$x = 36.4$

$\alpha = \mathbf{36.4\%}$

$\beta = 100 - 36.4 = \mathbf{63.6\%}$

Note: The proportion of the anomers in the *equilibrium* mixture of D-glucose is **36.4% α-D-glucopyranose and 63.6% β-D-glucopyranose**. Some of the other sugars in Table 6.1 may also undergo mutarotation and have differing amounts of anomers at equilibrium. The requirement to specify "equilibrium mixtures" is related to this phenomenon. [What is meant by *mutarotation* in relation to glucose anomers?]

7. (a) The value in Table 6.1 is for 1 g/mL in a 10 cm tube, or 100 g/100 mL, and is 66.5°

$\Rightarrow$ 26 g/L should give $\dfrac{26}{100} \times 66.5$ degrees

$= 17.29°$ (or 34.58° in a 20 cm tube)

This value (compared to 34.626°) checks out all right (allowing for rounding-off errors).

Note: The purpose of this question is to encourage students to always have an inquiring mind. As soon as you receive a piece of information that may have some relevance to something you have heard before, switch on your brain cells and ask: "How does this compare with what I learned before? Is it consistent?" [Your brain cells perform best if you feed information in a "consistent" manner.]

(b) 26 g/100 mL gives a rotation of 34.626.

24.632 g/100 mL, if 100% pure, should give a rotation of:

$\dfrac{24.632}{26} \times 34.626$ degrees

$= 32.8041$ degrees   (i)

As only 32.633 degrees is given, the sample is $\dfrac{32.633}{32.8041} \times 100$

$= \mathbf{99.48\%}$ **pure.**

Comments:

- No matter how technical some calculation may look, do not lose confidence in using your ratio method.
- The information that 34.626° is equated to 100°Z does not have to come into this calculation.
- Saying "*as only*" assures that extra bit of confidence — that the answer should come out "just short of 100%." Working out the expression (i) allows you to see that. Hence it was worked out, but note that extra significant figures have been retained. You can use $\dfrac{24.632}{26} \times$ 34.626 in place of the denominator, 32.8041, and get the same answer.

7
Enzyme Amounts

The amount of chemical and biochemical matter is measured in grams or moles. Because enzymes are functional entities, it is more useful to express amounts of enzymes in units that reflect their action capabilities or activities rather than merely their mass. There are several reasons for this approach:

1. It is likely that there are inactive (denatured) molecules in a given mass of enzymes.
2. Not all the relative molecular masses (M_r) of enzymes are known — to express them in moles. Often, two or more enzymes may be involved in a conversion. Crude systems (e.g., tissue homogenates) may be used as the enzyme source, and partially purified enzymes are often assayed for the very purpose of monitoring their purification.
3. Different enzymes have different catalytic capabilities (turnover numbers).
4. Some enzymes have more than one active site.

Common ways of expressing enzyme activity relate to rates of reactions catalysed by the enzymes. One is a *unit* of enzyme activity and another the *katal*. These and other relevant units are defined below.

One unit (U) of enzyme is the amount that will catalyse the transformation of one micromole of substrate per minute under specified conditions. The enzyme unit is spelt with a capital U if there is a likelihood of confusion with the word unit used in the ordinary sense. More recent definitions omit the word "amount" and emphasise enzyme activity. Thus, one unit of enzyme activity is the transformation or conversion of one micromole of substrate per minute. While various symbols have been used for an enzyme unit (e.g., U, EU, and u), it is recommended that the term be spelt out completely.

One katal (kat) of enzyme is the amount that will catalyse the transformation of one mole of substrate per second under specified conditions. Again, in a more recent version, one katal of enzyme activity involves the conversion of one mole of substrate in one second. As it is an excessively large unit, submultiples such as microkatals (µkat) or nanokatals (nkat) are often used.

Note that katal is defined in the SI units of moles and seconds. It is the preferred and recommended unit. Undergraduate students are advised to work out enzyme problems realistically (i.e., following first principles: how long was the incubation? do you want the answer in minutes or seconds? what amount of substrate was involved? was substrate in mmol or µmol units?) In this manner try to *arrive* at the appropriate units, such as moles substrate per second (which is katals). Do not blindly try to substitute information provided into conversion formulae.

Specific activity is an expression that relates activity to mass of protein. It must have an activity unit (units or katals) relating to a mass unit (e.g., units/mg of protein or units/g of protein). The SI preferred unit for specific activity is katals (or fractions of katals, such as µkat or nkat) per kilogram (e.g., µkat/kg). Where the relative molecular mass of an enzyme is known, molar activity (katals per mole of enzyme) is used.

Enzyme concentration in solutions is often expressed in units per millilitre or katals (or μkat or nkat) per litre (e.g., nkat/L).

Turnover number is the number of substrate molecules transformed (turned over) by one enzyme molecule per unit of time. The unit of time is usually a second or minute and needs to be specified.

Catalytic centre activity is turnover number *per active site* instead of *per molecule* (of enzyme).

Molecular activity is sometimes used to mean exactly the same as turnover number. The term *molecular activity* is particularly useful when comparison to catalytic centre activity is intended.

Questions

1. How many enzyme units are there in 1 katal?

2. How many nanokatals (nkat) are there in 1 unit?

3. 1 μg of an enzyme (M_r 160 000) converts 0.192 μmol/min. This enzyme is made up of four subunits. Calculate:
   (a) turnover number.
   (b) catalytic centre activity.

4. An experiment requires the use of 5×10^7 units of enzyme. The price for the enzyme is quoted for katal amounts. Allowing for a 20% excess for wastage, pipetting, and so forth, what approximate round number of katals of enzyme would you purchase?

5. Enzyme activity was monitored over the last two stages of an enzyme extraction-separation process. 0.4 g protein, from Stage 5, converted 46 mmol substrate in 10 sec, and 0.02 g protein, from Stage 6, converted 49 mmol in 30 sec.
   (a) Calculate the specific activities in units/g at each of the two steps.
   (b) Calculate % increase in activity obtained over the two steps.

6. 0.1 mg of a quaternary structured enzyme (M_r 486 000) with four active sites converts 48 μmol substrate/10 sec. Calculate:
   (a) turnover number (min^{-1}).
   (b) molecular activity (min^{-1}).
   (c) catalytic centre activity (min^{-1}).

7. An assay method in a clinical laboratory requires 1000 units of enzyme. From 0.5 katal available, how many assays can be performed. Assume no wastage. (If you consider it is not possible to perform even one assay, write "not possible.")

8. During an enzyme purification, 0.23 g of protein was dissolved in 5 mL buffer. Suppose 0.5 mL of this solution converted 4.7 mmol of substrate in 30 sec. Calculate specific activity as units/mg protein.

9. 10 μg of a pure enzyme (M_r 30 000) catalyses the conversion of 1.5 g of a substrate (M_r 44) to products in 5 minutes. Calculate:
   (a) turnover number (sec^{-1}).
   (b) units present in the experiment.
   (c) katals present in the experiment.

10. 1.0 μmol of an enzyme that contains five active sites per molecule converts 1.0 μmol substrate every 2 microseconds. Calculate catalytic centre activity (sec^{-1}).

11. An enzyme is available at a specific activity of 500 units/mg protein. For an assay, 10 μmol substrate need to be converted per minute. What mass (in mg) of protein would you take for this to be achieved?

12. 1.0 g wet weight of liver contains 50 units of an enzyme with a turnover number of 47 000. Calculate the intracellular concentration of the enzyme. Assume that 1 g weight of liver contains 0.8 mL water. Give your answer in μmol/L.

13. 0.046 g of a cell component, CC-12, was extracted from 3.0 g of a tissue and dissolved in 100 mL buffer. An enzyme, abicase, is known to be located in CC-12. 0.1 mL of the abicase-containing buffer converted 36.7 μmol of substrate in 5 minutes. Calculate:
   (a) units of abicase in 100 g CC-12.
   (b) units of abicase in 100 g tissue.

14. A purified hydrolytic enzyme (M$_r$ 150 000) is available with the description:
   Gives a V$_{max}$ of 3 200 units/mg protein under optimal conditions.
   (a) How many micromoles of substrate can be hydrolysed by 1 mg of protein in 1 minute?
   (b) How many nmol of enzyme are there in 1 mg?
   (c) Calculate its molecular activity (sec^{-1}).

Answers

1. 1 unit transforms 1 *μmol* substrate per *minute*.
   1 katal transforms 1 *mol* substrate per *second*.

   Note: Both time and quantity need to be changed.

   1 katal converts 1 mol of substrate per second.

   $\dfrac{1}{60}$ kat converts 1 mol/min*

   $\dfrac{1}{60}$ kat converts 10^6 μmol/min

   1 kat converts 60 × 10^6 = 6 × 10^7 μmol/min

   $$= 6 \times 10^7 \textbf{ units}$$

   * **Note:** $\frac{1}{60}$ kat is only a fraction of a kat, hence $\frac{1}{60}$ kat will work slower, taking a whole minute to do what 1 kat can do in 1 second.

2. 1 unit converts 1 μmol/min

   $$= \dfrac{1}{60} \text{ μmol/sec}$$

   $$= \dfrac{1}{60} \times 10^{-6} \text{ mol/sec}$$

$$= 0.01667 \times 10^{-6} \text{ mol/sec}$$

$$= 16.67 \text{ nmol/sec}$$

$$= \textbf{16.67 nkat}$$

3. (a) $\dfrac{1}{160\,000}$ μmol enzyme converts 0.192 μmol/min

$$\Rightarrow 1 \text{ μmol enzyme converts } 0.192 \times \frac{160\,000}{1}$$

$$= 30\,720 \text{ μmol/min}$$

Note: This figure is the number of μmol substrate converted by 1 μmol enzyme. It can also be taken as *moles substrate converted by 1 mole enzyme* or *molecules substrate converted by 1 molecule enzyme*. The *ratio* remains the same. Dividing both S and E by Avogadro's number is not necessary.

$\Rightarrow$ **30 720 is the turnover number**

(b) As the enzyme contains four subunits (i.e., four active sites), the turnover per active site is as follows:

$$= \frac{30\,720}{4}$$

$$= \textbf{7 680}$$

$$= \textbf{catalytic centre activity}$$

Note: As the question is given in μmol/minute, it is appropriate to leave the answers in min^{-1}.

4. The number of units needed $= \dfrac{120}{100} \times 5 \times 10^7$

$$= 6 \times 10^7 \text{ units}$$

$$= \textbf{1 katal}$$

Note: See answer to Question 7.1.

5. (a) **Stage 5:** 46 000 μmol/10 sec = 46 000 × 6 μmol/min

$$= 276\,000 \text{ units}$$

$$= 276\,000 \text{ units/0.4 g}$$

$$= 276\,000 \times \frac{1}{0.4} \text{ units/g}$$

$$= \textbf{690 000 units/g}$$

Stage 6: 49 000 μmol/30 sec = 98 000 units/0.02 g

$$= 4\,900\,000 \text{ units/g}$$

(b) % increase Stage 6 over Stage 5 $= \dfrac{4\,900\,000}{690\,000} \times 100$

$$= \mathbf{710\%}$$

6. (a) 48 µmol/10 sec = 288 µmol/min

100 µg converts 288 µmol/min

486 000 µg (= 1 µmol) converts $\dfrac{486\,000}{100} \times 288$ µmol/min

$$= \mathbf{1\,399\,680\ (turnover\ no.)}$$

(b) Molecular activity = Turnover No. = **1 399 680**

(c) Catalytic centre activity $= \dfrac{\text{Turnover No.}}{4}$

$$= \mathbf{344\,920}$$

Note: To keep close to the number of significant figures given in the question, it is better to express the answers as $\mathbf{1.40 \times 10^6}$ and $\mathbf{3.45 \times 10^5}$, respectively.

7. 0.5 kat = 3×10^7 units available. 1000 units are required for 1 assay:

$\Rightarrow\ 3 \times 107$ units required for $\dfrac{3 \times 10^7}{10^3}$

$$= \mathbf{3 \times 10^4\ or\ 30\,000\ assays}$$

8. 0.23 g protein/5 mL = 0.023 g protein/0.5 mL

4.7 mmol substrate converted/30 sec

= 9.4 mmol substrate converted/min

= 9 400 µmol substrate converted/min

= 9 400 units/0.023 g protein

= 9 400 units/23 mg protein

= **409 units/mg protein**

9. (a) 10 µg enzyme $= \dfrac{10}{30\,000}$ µmol enzyme

1.5 g substrate $= \dfrac{1.5}{44}$ mol $= \dfrac{1.5}{44} \times 10^6$ µmol converted/5 min

$$= \dfrac{1.5}{44} \times \dfrac{10^6}{5}\ \text{µmol converted/min}$$

$\dfrac{10}{30\,000}$ µmol E converts $\dfrac{1.5}{44} \times \dfrac{10^6}{5} \times \dfrac{1}{60}$ µmol substrate/sec

$\Rightarrow\ 1$ µmol E converts $\dfrac{30\,000}{10} \times \dfrac{1.5}{44} \times \dfrac{10^6}{5} \times \dfrac{1}{60}$

$$= 3.41 \times 10^5 \text{ μmol substrate/sec}$$

$$= \text{turnover no. (sec}^{-1})$$

(b) Units = amount that converts 1 μmol/min.

Here, $\dfrac{1.5}{44} \times \dfrac{10^6}{5}$ μmoles converted/min.

$$= 6.82 \times 10^3 \text{ units}$$

(c) 6×10^7 units = 1 kat

6.82×10^3 units $= \dfrac{6.82 \times 10^3}{6 \times 10^7}$

$$= 1.14 \times 10^{-4} \text{ kat}$$

$$= 114 \text{ μkat}$$

10. In 2×10^{-6} sec, 1 μmol converted

$\Rightarrow$ In 1 sec, $\dfrac{1}{2 \times 10^{-6}}$ μmol converted

This is by five active sites:

$\Rightarrow$ By one active site, $\dfrac{1}{2 \times 10^{-6}} \times \dfrac{1}{5}$

$$= 100\,000$$

$$= \text{catalytic centre activity (sec}^{-1})$$

11. 1 unit = 1 μmol substrate converted/min. We need 10 μmol substrate converted/min (i.e., need 10 units):

500 units in 1 mg

10 units in $\dfrac{10}{500} \times 1$

$$= 0.02 \text{ mg}$$

12. 50 units/0.80 mL $= \dfrac{50}{0.8}$ units/mL

1 unit = amount that converts 1 μmol substrate/min

$\Rightarrow \dfrac{50}{0.8}$ units = amount that converts $\dfrac{50}{0.8}$ μmol substrate/min

Turnover no. is 47 000 (i.e., 47 000 μmol substrate converted by 1 μmol enzyme):

$\Rightarrow \dfrac{50}{0.8}$ μmol converted by $\dfrac{50}{0.8} \times \dfrac{1}{47\,000}$ μmol enzyme

$$= 0.001\,33 \text{ μmol enzyme/mL}$$

$$= 1.33 \text{ μmol/L}$$

Note: Students may wish to consider this problem in an "everyday situation" as follows:

An amount of money (x) buys $\dfrac{50}{0.8} = 62.5$ articles

47 000 articles can be bought for $1.00

62.5 articles can be bought for $\dfrac{62.5}{47\,000} \times 1$ dollars

= **$0.001 33**

This is the amount of money x. (No, don't change it to µ$!)

13. (a) 1 unit = 1 µmol substrate converted/min

   36.7 µmoles/5 min = 7.34 units/0.1 mL

   $\qquad = 7.34 \times 10^3$ units in 100 mL or 0.046 g

   $\qquad = 7.34 \times 10^3 \times \dfrac{100}{0.046}$ units/100 g CC-12

   $\qquad$ = **1.60×10^7 units/100 g CC-12**

   (b) 0.046 g CC-12 = 3 g tissue

   So, 7.34×10^3 units in 3 g tissue

   $= 7.34 \times 10^3 \times \dfrac{100}{3}$ units/100 g tissue

   = **2.45×10^5 units/100 g tissue**

   Note: Expect answer (b) to be smaller than answer (a) because answer (b) is considered in relation to the *whole* tissue.

14. (a) 1 unit converts 1 µmol substrate/min

   3 200 units convert **3 200 µmol substrate/min**

   (b) 1 g contains $\dfrac{1}{150\,000}$ mol enzyme

   1 mg contains $\dfrac{1}{150\,000} \times 10^{-3} \times 10^9$ nmol

   $\qquad$ = **6.67 nmol**

   (c) Molecular activity = Turnover no.

   6.67 nmol enzyme converts $\dfrac{3200}{60}$ µmol substrate/sec

   $\Rightarrow$ 1 nmol enzyme converts $\dfrac{1}{6.67} \times \dfrac{3200}{60} \times 10^3$ nmol substrate/sec

   $\qquad$ = 8 000 nmol substrate/sec

Molecular activity = 8 000

Note: In the final step of the calculation, **n**mol enzyme has been related to **n**mol substrate, so this value becomes the turnover number or the molecular activity. See comments following the answer to Question 7.3(a).

8
Enzyme Kinetics

It is beyond the scope of this book to go into an advanced treatment of enzyme kinetics. Perhaps what is most important is to point out the role of enzyme kinetics — something that is often lost by student and teacher when absorbed in with the quantitative aspects of enzyme kinetics. As the word *kinetics* suggests, this topic is clearly a quantitative one.

Enzyme kinetics deals with the rates of enzyme-catalysed reactions. Study of the rates of reaction and how they change in response to changes in experimental parameters adds a great deal to our understanding of enzyme mechanisms. The kinetic approach is the oldest approach and continues to remain the most important even today. The reason is simple: we cannot look directly at the active sites at the level of atoms; we therefore have to draw conclusions from other practical, usually quantitative, results.

Kinetic data is often plotted on graphs. Information on enzyme characteristics (such as K_m and V_{max}), mechanisms (ping-pong or sequential binding of substrates), the nature of inhibitors (competitive or noncompetitive), and the nature of enzymes (simple or allosteric) can all be obtained from kinetic graphs. This chapter deals with simple enzyme kinetics, including the presentation of kinetic data on graphs and their interpretation. As presentation of graphs and other figures is an important part of scientific report writing, the rules or conventions relating to their presentation will also be reviewed in this chapter.

Summary of Enzyme Kinetics — the Equations and the Graphs

1. For detailed discussions, students are referred to the general biochemistry textbooks listed in the bibliography.
2. The modern derivation of the **Michaelis–Menten Equation** for the assumed enzyme catalysed process starts with:

$$\text{E} \; + \; \underset{k_{-1}}{\overset{k_1}{\rightleftharpoons}} \; \text{ES} \; \xrightarrow{k_2} \; \text{E} \; + \; \text{P} \tag{i}$$

- A great deal of the derivation hinges on the ES, the enzyme-substrate complex.
- Very soon after the start of the reaction (at the so-called *initial* conditions), the concentration of the product [P] is negligible, hence there is no reverse reaction; accordingly, k_{-2} is ignored.

$$\text{ES} \; \xleftarrow[k_{-2}]{} \; \text{E} \; + \; \text{P} \tag{ii}$$

3. In the Michaelis–Menten derivation, note very carefully how *they*, and subsequently every-body else, collected the kinetic constants and the concentration terms.

$$\frac{([E_t] - [ES])[S]}{[ES]} = \frac{k_2 + k_{-1}}{k_1} = K_m \tag{iii}$$

Where E_t (or some textbooks have it as E_o) represents starting/initial/total amount of enzyme.

The terms were collected in such a manner that k_2 and k_{-1} appeared as the numerator and *then* the **K_m, the Michaelis–Menten constant,** assigned to the fraction. Both k_2 and k_{-1} represent the breakdown of ES. Hence, the greater the K_m, the greater the instability of ES or lack of affinity of E for S; it does not mean a faster enzyme catalysed reaction (produc-ing P).

Most textbooks include a discussion on the *significance* of K_m. Do be aware that K_m (pos-sibly because it is based on assumed processes and their rate constants) is not an easy con-cept to fully understand or to apply universally. While the statement about "the instability of ES" is true, it is a misconception that K_m directly indicates the (inverse) affinity of the enzyme for the substrate. (This will only be the case if k_2 is negligible compared to k_{-1}. In this case, the enzyme would display very poor catalytic activity, which in most cases is not so.) When K_m values are compared, some useful points are observed. A small K_m means only small amounts of substrate are needed for the enzyme to be fully saturated and reach maximum velocity. Conversely, a large K_m would mean high substrate concentrations are needed. Another point: where an enzyme acts on several (closely related) substrates, com-parison of the K_m values would indicate which is the "most-suited" substrate — frequently (not always), it is the enzyme's natural substrate.

4. The last equation (iii) is not the Michaelis–Menten equation. Due partly to the fact the [ES] cannot be measured directly, the last equation has to be reexpressed to include more measurable parameters, v_o (**initial velocity**) and V_{max} (**maximum velocity**), to give the **Michaelis–Menten equation**:

$$v_o = \frac{V_{max}[S]}{K_m + [S]} \tag{iv}$$

The term v_o worries a lot of students, but it simply means the measured velocity — measured by substrate disappearance or product formation. This can be done under any condition, low substrate concentration, high substrate concentration, wrong pH, and so forth. The *initial* means *early in time*, before significant [P] has formed, which would slow down the reaction (i).

V_{max}, the maximum velocity, is obtained from a Michaelis–Menten plot (see Figure 8.1). This V_{max} is for a given or fixed amount of enzyme. Naturally, a different V_{max} will be obtained for a different amount of enzyme. Twice $[E_t]$ will give a new V_{max} (twice that of the former). In the Michaelis–Menten plot (and other plots, such as the Lineweaver–Burk plot), the amount of enzyme is fixed; it is the substrate concentration that varies (and the resulting velocity is plotted for each substrate concentration).

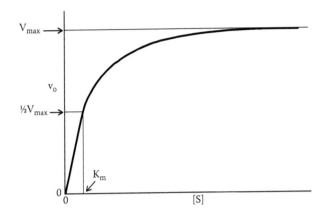

Figure 8.1   The Michaelis–Menten plot.

5. An interesting mathematical trick can be performed by arbitrarily considering what happens to equation (iv) when v_0 is exactly one-half the value of V_{max}. It is with this trick that we get (by substituting and solving):

$$K_m = [S], \text{ when } v_0 = \frac{1}{2} V_{max}$$

on the graph. It is a simple and practical way of working out the value of K_m. Note, however, that just saying that K_m corresponds to the substrate concentration at exactly half the maximum velocity does not make the meaning of K_m any clearer.

6. The unit for K_m is molar (mol/L) but can be written as fractions of mol/L, such as mmol/L, μmol/L, etc.

7. The Michaelis–Menten plot is the direct plot of rate (or velocity) against [S]. Rate must be on the y-axis and [S] on x-axis (see Figure 8.1). For general rules on the presentation of graphs and other figures, see Box 8.1.

8. An important transformation of the Michaelis–Menten equation is the **Lineweaver–Burk equation,** obtained basically by inverting equation (iv) and importantly, making it to match the equation of a straight line, $y = mx + c$.

$$\frac{1}{v_0} = \frac{K_m}{V_{max}} \frac{1}{[S]} + \frac{1}{V_{max}}$$

$$y \;=\; m \quad x \;+\; c$$

Thus, as shown in Figure 8.2, $\frac{1}{v_0}$ goes on the y-axis, $\frac{1}{[S]}$ on the x-axis, the intercept on the y-axis $= \frac{1}{V_{max}}$, and the intercept on the x-axis is $-\frac{1}{K_m}$. The slope, m, equals $\frac{K_m}{V_{max}}$.

9. Another transformation of the Michaelis–Menten equation is the **Hofstee–Eadie equation**.

$$v_0 = V_{max} - K_m \frac{v_0}{[S]}$$

$$y \;=\; c \quad - m \quad x$$

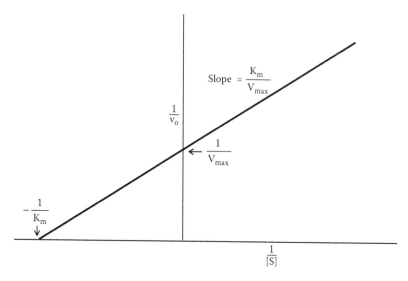

Figure 8.2   The double-reciprocal Lineweaver–Burk plot.

This transformation again matches the equation of a straight line, albeit somewhat back-to-front. The Hofstee–Eadie plot, with the information that can be obtained from it, is shown in Figure 8.3.

10. There are other transformations giving further equations and plots. What is the point of the different equations and plots? Basically, each has some particular advantage in analysing enzyme kinetic data. The more advanced student is expected to know the practical advantages of each.

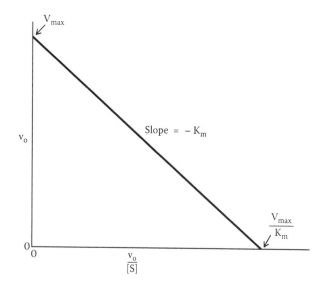

Figure 8.3   The Hofstee–Eadie plot.

Box 8.1   Presentation of Figures

Graphic analysis is important not only in enzyme kinetics but also in a number of other areas of quantitative biochemistry. The rules or conventions generally followed in the presentation of graphs and other illustrations (collectively called *figures*) are summarised in this box.

1. Figures should have clear titles and be numbered. When there is more than one figure in a report, label them in sequence starting from "Figure 1."
2. Chemical formulae and equations should not be set apart from the text and should not be called figures. However, if several formulae or equations are grouped together, then the groups may be called figures or even tables (see Table 5.1). Biochemical pathways or sets of illustrative chemical structures showing, for example, sequential deprotonation, should be called figures (see Figure 5.4).
3. Figures (usually graphs) in undergraduate reports must be located in the report close to where they are referred to. Generally, provide a table containing relevant results as well as the graph (tabled results are unnecessary in published papers).
4. Each figure must have a descriptive caption. Place the caption *below* the figure. The caption should be complete enough to stand on its own without any need to refer to the text. A caption can simply be a title (e.g., "Figure 2. Absorbance at 500 nm *vs* glucose concentration").
5. In undergraduate reports, label different graphs (curves) on the same figure *directly* (although different symbols are used in published papers — in which case, explanation of the symbols should be included in the caption). A legend is an explanatory statement that accompanies a figure, including explaining the symbols. Some editors of journals use the terms *legend* and *caption* interchangeably.
6. Choose sizes of graph paper, scales, and scale markings carefully. For most undergraduate reports, the graph paper should be about half an A4 sheet, and the graph should fill at least 70% of it. Trim and paste the graph paper flat; no folding over is allowed.
7. Scale markings should be "round figures" only, evenly spaced. Try to have between three and eight values. For example:

pH	2	4	6	8	10	
or	0	2.5	7.5	10		
Wavelength (nm)	300	350	400	450	500	

Do not show values obtained from an experiment (e.g., 2.76) on a scale.
8. Choose scales carefully in relation to the results. Generally, try to show a "normal" response by a line of gradient 1 (45°). Do not exaggerate the scale to "sell" your results or as a result of poor judgment.
9. Where the curves are related, or being compared, plot them on the same graph. If several graphs are related, use the scales consistently; otherwise comparison is difficult.
10. Use the *x*-axis for the independent variable and the *y*-axis for the dependent variable. They should not be interchanged when dealing with scientific data. Conventions followed for most biochemical work require that time, temperature, pH, and substrate concentration go on the *x*-axis. Sadly, a number of textbooks are amiss on this. A bell-shaped response, resulting from the effect of pH on an enzyme-catalysed reaction, must have the bell sitting on its base and not lying on its side.

11. Label the *x*-axis horizontally; the *y*-axis may be labeled "running up the page." Place labels, including units of measurement, centrally.
12. If figures have to be turned 90° to the writing page, the base (caption) must be on the right-hand edge of the page.
13. Do not draw arrowheads on ends of axes except in mathematical contexts where their presence is relevant and necessary.
14. Do not draw guidelines (perpendicular to axes) for points plotted on graph.
15. Include only meaningful limits of the scale(s) — often starting from zero. A break in the axis must be shown if appropriate. Values such as pH and wavelength of electromagnetic radiation (light) need not start from zero.
16. Treat the point (0,0) when "set" on instrument like any other point in drawing the line of best fit. Do not force the line to pass through zero.
17. Do not draw one-point graphs purely to create an impression in your reports. These graphs assume linear response, in which case the ratio method calculation (comparing to a standard value) is more accurate.
18. Graphs do lose mathematical accuracy although they provide pictorial effect. Unless graphs "average" values (e.g., line of best fit), it may be more accurate to go to original values for calculations. (see Chapter 5 and the answer to Question 8.6 for further comments on the pictorial nature of graphs and computer-aided graphic analysis.)

Questions

1. Tubes containing varying concentrations of substrate were incubated with 1.0 µg enzyme (M_r 40 000) in a final volume of 4.0 mL. A 0.5 mL reaction mixture was withdrawn at 30-second intervals and assayed for the product formed (see Table 8.1). Estimate the initial velocities (v_o) as µmol product formed per minute for each substrate concentration.

2. Using the values from Question 1, plot a Michaelis–Menten curve and a Lineweaver–Burk plot and determine V_{max} and K_m. Before any graph-plotting, estimate (and write down) what the V_{max} and K_m should be by simply inspecting the figures.

Table 8.1   Data for Question 8.1

Substrate (mmol/L)	Product (µmol) at:			
	0 sec	30 sec	60 sec	90 sec
2	0	0.44	0.89	1.33
3	0	0.57	1.15	1.71
5	0	0.73	1.45	2.17
15	0	1.11	2.20	3.32
40	0	1.35	2.71	4.06
100	0	1.40	2.73	4.02
200	0	1.44	2.79	4.05

Table 8.2   Data for Question 8.3

Serine Concentration (µmol/L)	HO-Pyr Formed (µmol/min)
1.68×10^{-3}	0.106
4.2×10^{-3}	0.200
8.4×10^{-3}	0.250
16.8×10^{-3}	0.310
84.0×10^{-3}	0.320

3. A bacterial enzyme deaminates serine as follows:

$$\text{Serine} \longrightarrow \text{Hydroxypyruvic Acid} + NH_3$$

The initial rate of the reaction was measured at various concentrations of serine at 25° with the enzyme concentration held constant (Table 8.2).

(a) Decide whether you would use a Michaelis–Menten plot or a Lineweaver–Burk plot to determine graphically the V_{max} and K_m.

(b) Plot the graph you have decided on, and determine graphically the V_{max} and K_m.

4. An enzyme-catalysed reaction was found to be inhibited by compound C. The following data were obtained for initial velocities at various substrate concentrations both in the presence (0.1 mmol/L C) and absence of the inhibitor (see Table 8.3).

(a) Determine graphically the K_m values with and without the inhibitor.

(b) Determine graphically the V_{max} values with and without the inhibitor.

(c) What type (competitive or noncompetitive) of inhibitor is compound C? Why?

5. Five tubes were set up, each containing 1 µg enzyme (K_m 2.5×10^{-4} mol/L) and 0.2 µmol of a competitive inhibitor. The substrate concentration was varied, as shown in Table 8.4. The tubes were incubated at 37° and initial reaction rates were determined. Determine graphically the velocity of the *uninhibited* reaction at a substrate concentration of 0.50 mmol/L.

6. Using the values from Question 8.3, generate a Hofstee–Eadie plot and determine V_{max} and K_m. Make brief comments comparing the Lineweaver–Burk plot (Question 8.3) and this plot.

Table 8.3   Data for Question 8.4

Substrate (mmol/L)	Rate of Disappearance of S = Initial Velocity (µmol/min)	
	No Inhibitor	With Inhibitor
1.33	10.5	4.3
1.67	12.7	5.3
2.22	15.9	6.9
3.33	21.7	10.0
6.67	33.3	17.5

Table 8.4 Data for Question 8.5	
Substrate (mmol/L)	Reaction Rate with Inhibitor (μmol S converted min^{-1})
0.10	18
0.15	24
0.20	30
0.50	51
0.75	63

7. Sketch the following curves for an enzyme obeying Michaelis–Menten kinetics. Label all axes carefully.
   (a) v *vs* [S]
   (b) v *vs* [E]
   (c) v *vs* pH
   (d) v *vs* temperature
   (e) v *vs* time for $[S] \gg K_m$
   (f) [P] *vs* time for $[S] \gg K_m$
   (g) [P] *vs* time for $[S] \ll K_m$

Answers

1. Examine the figures in Table 8.1. From 2 mmol/L to 40 mmol/L, the μmol produced at 30, 60, and 90 seconds double and triple, indicating a direct increase or linear response (and no slowing down). We could waste time drawing little graphs but simple inspection is going to give us the best possible figures. Thus, the initial velocities for the 2, 3, 5, 15, and 40 mmol/L substrate concentrations respectively are:

    0.89   1.15   1.45   2.20   2.71 μmol/min

    What about the initial velocities for the 100 and 200 mmol/L concentrations? Carefully examine the 30-sec, 60-sec, and 90-sec results for these two concentrations. For both concentrations, the doubling and tripling responses are not present. As substrate is used up at a fast rate, substrate becomes limiting *later in time*. The rate of reaction is slowing down — we must not use the rate here as representing *initial velocity*. What then are the initial velocities for these concentrations? As we do not have any values at less-than-30-seconds, we have to use the 30-second figures to represent initial velocity.

    For 100 mmol/L substrate, 2 × 1.40 = **2.80 μmol/min**
    For 200 mmol/L substrate, 2 × 1.44 = **2.88 μmol/min**

    What else can we say from the figures? Note initial velocities (which are represented by the 60-second figures, except for the two above) increase progressively and correspondingly to the increase in substrate concentrations; even slightly to 100 mmol/L [S] but hardly above the 100 to 200 mmol/L [S]. This would mean "all the available enzyme" or "the enzyme that is present" (1 μg) is *saturated* with substrate. (The word *saturated* in this context means every enzyme molecule is working as fast as it can and has to become free before it can perform another catalysed reaction — that we can *see* when we measure the product formed.) Falling back free is of no relevance here as we are monitoring the production of product. (If you wish to get really technical, in all these enzyme-catalysed reactions where we observe the *product formed*, we are indeed monitoring:

$$ES \longrightarrow E + P$$

part of the Michaelis–Menten hypothesized reaction.)

(Question 8.1 simply asked you to figure out the *initial* velocities. Question 8.2 takes you into the *kinetics* plots.)

2. The initial velocities, obtained from Question 8.1, are shown in Table 8.5. The V_{max} should be 2.88 or higher. Agreed? The K_m should be around 5 mmol/L. Why? Because it should be the substrate concentration when $v_o = \frac{1}{2} V_{max}$. So whatever you do with the values — finding reciprocals, plotting graphs, etc. — make sure you come back to about 5 mmol/L (note the unit *is* mmol/L).

The Michaelis–Menten plot (see Figure 8.4) is a direct plot, whereas we need reciprocals for the Lineweaver–Burk plot (see Table 8.6).

Note it is hard to estimate V_{max} on the Michaelis–Menten plot because the curve does not completely flatten out, consequently we cannot estimate $\frac{1}{2} V_{max}$ and, therefore K_m, accurately. The Lineweaver–Burk plots allow more importance to be placed on the lower [S] values and velocities (generally considered more reliable).

The $\dfrac{1}{V_{max}}$ read off the graph (see Figure 8.5) is 0.342

$\Rightarrow$ **V_{max} = 2.92 µmol/min**.

The $-\dfrac{1}{K_m}$ read off the graph is −0.22

$\Rightarrow$ **K_m = 4.5 mmol/L**

Note: Both answers match our earlier predictions.

Table 8.5 Data for Question 8.2 (Initial Velocities from Question 8.1)	
S (mmol/L)	v_o (µmol/min)
2	0.89
3	1.15
5	1.45
15	2.20
40	2.71
100	2.80
200	2.88

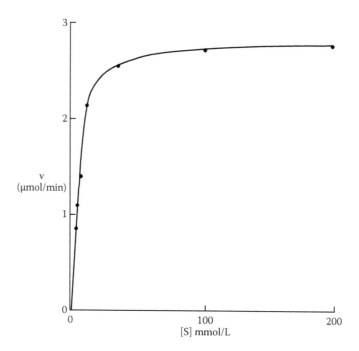

Figure 8.4   Michaelis–Menten plot for Question 8.2.

3. (a) Note from the v values given in the question, V_{max} has not been reached. Therefore (and for greater accuracy), the Lineweaver–Burk plot (and not the Michaelis–Menten plot) has to be used.

  (b) For the Lineweaver–Burk plot, we need reciprocals of [S] and v (see Table 8.7).

  From the Lineweaver–Burk plot (see Figure 8.6):

$$\frac{1}{V_{max}} = 2.64$$

$$\Rightarrow V_{max} = 0.379 \ \mu mol/min$$

Table 8.6 Reciprocals for Question 8.2	
$\dfrac{1}{[S]}$	$\dfrac{1}{v_0}$
0.50	1.124
0.33	0.870
0.20	0.690
0.067	0.452
0.025	0.370
0.010	0.357
0.005	0.347

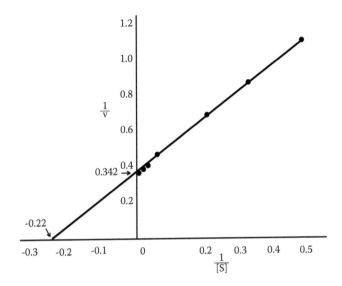

Figure 8.5   Lineweaver–Burk plot for Question 8.2.

Note: The answer is greater than the highest figure given the table (and of the same magnitude).

$$-\frac{1}{K_m} = -0.240$$

$$\Rightarrow K_m = 4.17 \text{ (the unit is \textbf{mmol/L})}$$

Note: We can approximate the answer. V_{max} is 0.379, half that is 0.19, and the [S] corresponding to 0.19 is slightly less than 4.2.

General Comments:
Finding reciprocals of complex numbers can be difficult enough and can be more confusing when we have to find the reciprocal of reciprocals (as we have to do to get back to K_m and V_{max} from the Lineweaver–Burk graphs). The advice is to use simple numbers wherever possible: 1.68×10^{-3} mol/L becomes 1.68 mmol/L, but give final answers in the same units as in the question; give the final answer as 4.23×10^{-3} mol/L (not 4.23 mmol/L). Reciprocals of numbers such as 1.68×10^{-3} or 1.68×10^{-5} mol/L will just bring in too many noughts; avoid them. Go for mmol/L, μmol/L, and the like.

Table 8.7   Reciprocals for Question 8.3

Serine (mmol/L)	$\dfrac{1}{[S]}$ (mmol/L)	HO-Pyr Produced (μmoles/min)	$\dfrac{1}{v}$ (μmol/min)
1.68	0.595	0.106	9.43
4.20	0.238	0.200	5.00
8.40	0.119	0.250	4.00
16.8	0.060	0.310	3.23
84.0	0.012	0.320	3.12

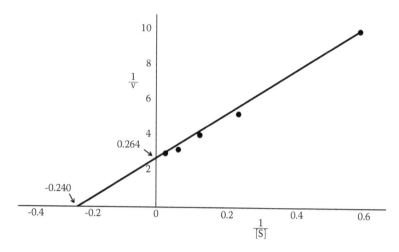

Figure 8.6   Lineweaver–Burk plot for Question 8.3.

The units for $\frac{1}{[S]}$ and $\frac{1}{v}$ are not easy and are cumbersome to write down. (Recall in Chapter 1, you were discouraged from using negative exponentials with units.) The advice is to leave them alone; just note down (and be careful) what units are used for [S] and v. For example, in $\frac{1}{[S](mmol/L)}$, the (mmol/L) indicates that the [S] is in mmol/L.

In all graph plotting, but in particular with the Lineweaver–Burk plot, choose sensible scales for the axes. Note that you will need to extrapolate the graph to the LHS to obtain K_m; hence the y-axis has to be centrally placed. Negative $\frac{1}{K_m}$ appears on the LHS where the graph crosses the negative x-axis. Care is needed in reading negative scales. They have to be read from right to left. Why not remind yourself by clearly including a negative sign, as shown in Figures 8.5 and 8.6.

4. For Lineweaver–Burk plots, reciprocals are needed (see Table 8.8). The Lineweaver–Burk plots are shown in Figure 8.7. Note the even spread of points. The trick in designing experiments of this kind is to choose a regular spread of *reciprocals* of S, and from them obtain the set of S that is then used in the experiment.

Note: Please read the comments given with the answer to Question 8.3, if you have not already.

Table 8.8   Reciprocals for Question 8.4

| | | Initial Velocity (µmol/min) | | | |
| | | Without Inhibitor | | With Inhibitor | |
S (mol/L)	$\frac{1}{[S]}$	v	$\frac{1}{v}$	v	$\frac{1}{v}$
1.33	0.75	10.5	0.095	4.3	0.233
1.67	0.60	12.7	0.079	5.3	0.189
2.22	0.45	15.9	0.063	6.9	0.145
3.33	0.30	21.7	0.046	10.0	0.100
6.67	0.15	33.3	0.030	17.5	0.057

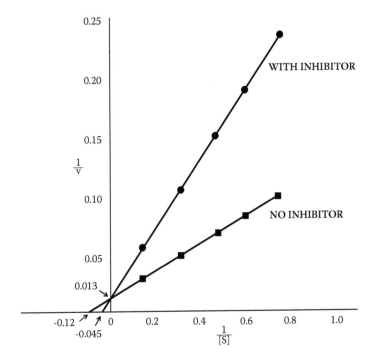

Figure 8.7   Lineweaver–Burk plots for Question 8.4.

(a) K_m value without inhibitor

$$-\frac{1}{K_m} = -0.12$$

$$\Rightarrow K_m = 8.3 \text{ mmol/L}$$

K_m value with inhibitor (apparent K_m)

$$-\frac{1}{K_m} = -0.045$$

K_m = 22 mmol/L

Note: Yes, it is meant to be higher than the one without the inhibitor.

(b) Let us consider V_{max} and uninhibited plot only for the moment.

$$\frac{1}{V_{max}} = 0.013$$

$$\Rightarrow V_{max} = 77 \text{ µmol/min}$$

Note that the value 77 µmol/min is feasible because for the S values given there is no obvious slowing down of the velocities. (Table 8.8, we expected something much greater than the highest value, 33.3 µmol/min.) What is interesting (and an excellent evaluation of our answers) is that $\frac{1}{2}$ (77) = 34 µmol/min, and the [S] corresponding to about 34 µmol/min is about 7 or 8 mmol/L. (Table 8.8, again higher than the highest S value, 6.67 mmol/L. The answer we got is 8.3 mmol/L. *How exciting!*)

(What else can we say? Looks like all the values we have in Table 8.8 must be from the very low part of the Michaelis–Menten plot — well before even V_{max} was reached. In the comments following the answer to Question 8.6, we conclude that this is a good area to work in.)

(c) Now for the fact that the V_{max} for both plots (inhibited and uninhibited) are the same. This is kinetic evidence that the inhibition is **competitive**. (Please refer to textbooks if you are not familiar with the reasons for this deduction.) The fact that K_m is altered (the apparent K_m *increased*) also supports the conclusion that C is a competitive inhibitor.

5. A Lineweaver–Burk plot is generated using the inhibited rate values (see Figure 8.8). We are told that the inhibition is competitive, so the V_{max} will be coincidental on the two plots. We are given the K_m for the enzyme. The reciprocal of the K_m is located on the x-axis (the $\frac{1}{[S]}$-axis). K_m given as 2.5×10^{-4} mol/L = 0.25 mmol/L. $-\frac{1}{K_m}$ is -4 (on the $\frac{1}{[S]}$-axis). This point and $\frac{1}{V_{max}}$ are joined and extrapolated as shown on the graph. The velocity (actually $\frac{1}{v}$) corresponding to 0.50 mmol/L (actually, $\frac{1}{0.50} = 2$ on the $\frac{1}{[S]}$-axis) is read off the curve (and worked back to v).
From the graph:

$$\frac{1}{v} = 0.016$$

$$v = \textbf{62.5 μmol/min}$$

Note: Naturally, we expect an answer greater than 51 μmol/min (which is the inhibited rate).

6. For the Hofstee–Eadie plot, we need [S], v, and $\frac{v}{[S]}$ (see Table 8.9). The graph is plotted v (on the y-axis) vs $\frac{v}{[S]}$ (on the x-axis), as shown in Figure 8.9.

Note: Do not worry about creating units for the fraction $\frac{v}{[S]}$. Remember you must use simple numbers for S and v (if not provided as such). On the graph, the scientific notation

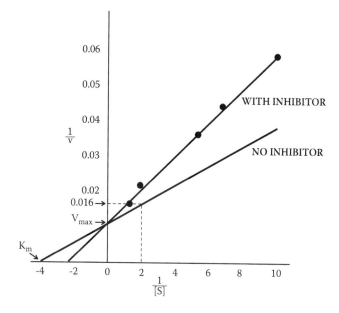

Figure 8.8   Lineweaver–Burk plots for Question 8.5.

Table 8.9   Data for Hofstee–Eadie Plot (Question 8.6)

[S] (mmol/L)	v (μmol/min)	$\dfrac{v}{[S]}$
1.68	0.106	$0.0631 = 6.31 \times 10^{-2}$
4.2	0.200	$0.0476 = 4.76 \times 10^{-2}$
8.4	0.250	$0.0298 = 2.98 \times 10^{-2}$
16.8	0.310	$0.0185 = 1.85 \times 10^{-2}$
84.0	0.320	$0.0119 = 1.19 \times 10^{-2}$

form is possibly the best way of showing these values on the axis — this avoids marking too many zeros. In this case, it does not matter too much, but avoid markings such as 0.00001 and 0.00002 on graphs.

$$V_{max} = \text{intercept on } y\text{-axis}$$

$$= \textbf{0.380 μmol/min}$$

$$-K_m = \text{slope (Note that the gradient is negative.)}$$

$$= -\frac{y - \text{intercept}}{x - \text{intercept}}$$

$$= -\frac{0.380}{0.095}$$

$$= -4.0$$

$$K_m = 4.0$$

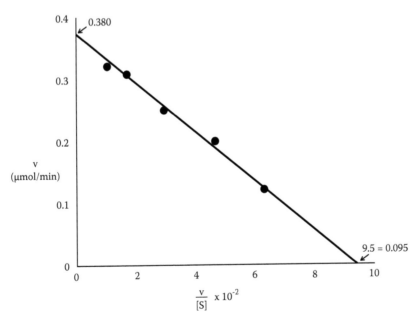

Figure 8.9   Hofstee–Eadie plot for Question 8.6.

What about the unit for K_m? We expect it to be mmol/L. Using the factor-label cancellation method, we can prove this:

$$-\frac{0.380\ \mu mol/min}{0.095\ \dfrac{\mu mol/min}{mmol/L}}\ ^* = -\textbf{4.0 mmol/L}$$

*Note: Do you follow this? On cancelling the "μmol/ min," the expression becomes $\dfrac{-\frac{1}{1}}{\frac{1}{mmol/L}}$, which is mmol/L.

Comments on Questions

1. Both the V_{max} and K_m compare favourably to those obtained from the Lineweaver–Burk plot (Question 8.3). Admittedly, this is as a result of very careful plotting and drawing lines of best fit.

2. As both the Hofstee–Eadie and the Lineweaver–Burk plots depend critically on the line of best fit, the wider the range of points on the graph, the more reliable the fit. It is no good having a large number of points all compacted together. The Lineweaver–Burk plot would produce better results if we had a large range of values at low substrate concentrations and low velocities (we saw that in Question 8.4). (The higher values become more compacted close to V_{max} — maybe that is a good thing for estimating V_{max}). A disadvantage is that a long extrapolation is often required to estimate K_m, causing uncertainty of the result. One of the problems with Hofstee–Eadie plots is in using the combined value $\frac{v}{[S]}$. Errors in the practical determination of either v or [S] can cause serious distortions on $\frac{v}{[S]}$. As mentioned earlier, both plots benefit from having a wide range of values. In practical situations, this is not always attainable for numerous reasons that vary from enzyme to enzyme.

   In research, it is expected these days to perform computer analysis of data. Computers can be used to perform different graphic analysis of data in incredibly short times and avoid the human errors and influence involved in plotting, reading values, drawing lines of best fit, and extrapolation. Computer programs allow lines of best fit to be statistically computed, providing regression coefficients and other statistical test values indicating the accuracy of the estimates.

3. Sketches for Question 7 are shown in Figure 8.10.

Comments on Answers

1. Note that there is a difference between plotting a graph and sketching a curve. The latter demands more scientific understanding. The former is simply tedious. In sketches, careful attention must be paid to turning points, intersections with axes, symmetry, sharp decline, and so forth. Furthermore, critical values must be shown (and labeled) on the sketches (like the value where it crosses the y-axis, maximum or minimum values).

2. In Figure 8.10(a), the top of the curve should be horizontal. In (c), the bell-shaped response should be shown with both sides symmetrical. In (d), the ascending part should be gradual while the descending part should show a sharp decline in activity when temperature is raised beyond a point when considerable denaturation sets in. If you are uncertain for other scientific implications of sketches (a) to (d), please refer to textbooks.

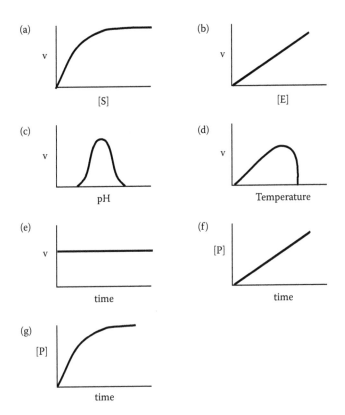

Figure 8.10   Sketches for Question 8.7.

3. It is important that students understand the meaning of $[S] \gg K_m$ and $[S] \ll K_m$. Terms such as "high substrate concentration" and "low substrate concentration" are meaningless because there is no reference given. What is a high substrate concentration in one enzyme system may not be so in another. In enzyme kinetics and Michaelis–Menten work, the terms *high* and *low* are made with reference to K_m (see Figure 8.1). *High* means a substrate concentration higher than K_m, and $[S] \gg K_m$ means definitely we are dealing with substrate concentrations at the plateau of the curve (where zero order kinetics apply). The statement $[S] \ll K_m$ means at low substrate concentrations, at the ascending part of the Michaelis–Menten curve where first order kinetics apply. Hence, for (e), the velocity is constant with respect to time because the given amount of enzyme is working as fast as it can. If the velocity is constant, then the amount of product produced will increase constantly with respect to time indefinitely (f). On the other hand, when substrate is limiting when $[S] \ll K_m$, then the amount of product will only increase for a limited period and stay at that level when substrate is exhausted (g).

9

Colorimetry and Spectrophotometry

Quantitative estimation of biological substances is understandably a very important part of the science of biochemistry. In chemistry, students would have dealt with analytical techniques involving titrations (titrimetric analysis) or weighing (gravimetric analysis). The most important and extensively used assay technique in biochemistry is **spectrophotometry.**

Photometry, the generic name that means light-measurements, includes absorption spectrophotometry (a subset of which is **colorimetry**), **fluorometry, flame photometry,** and **atomic absorption.** Similar principles and calculations are involved in all these techniques. This chapter will deal mainly with colorimetry and spectrophotometry.

Spectrophotometric measurements are ordinarily made in the spectral range 220–800 nm.[*] This range is subdivided into the visible range above 380 nm and the ultraviolet (UV) range below 380 nm. The infrared (IR) region that extends above 800 nm finds little application in biochemical analysis. It is used in organic chemistry to identify organic groups. Colorimetric analyses are made in the visible range; spectrophotometric analysis usually implies analysis in the UV range, although it could include analyses made in the visible range as well.

Both light transmission (meaning amount of emergent light after passing through a colorimeter tube or cuvette containing the light absorbing solution) and absorbance (meaning the absorbance by the solution or depth of color of the solution) may be used as the indicator of the concentration of the solution in the cuvette. The latter, absorbance, or optical density (OD), is more frequently used in biochemistry. This is because there is a linear relationship between concentration of solutions (this is what we are after) and absorbance. The underlying law relating to all spectrophotometric calculations is the Beer–Lambert Law, which simply states that the absorbance of the solution is directly proportional to the concentration and the length of light path through the solution (the width of the cuvette):

$$\text{Log}_{10} \frac{I_o}{I_e} = Kcl$$

where

I_o = intensity of incident light
I_e = intensity of emergent light
K = molar absorbancy index or extinction coefficient (ε is also used instead of K)
c = concentration of the solution
l = length of light path through the solution (usually 1 cm)

---

[*] In biochemistry, nm (10^{-9} m) is the preferred unit for wavelength. The older, Å Angstrom unit (10^{-10} m), is no longer used.

The term $\text{Log}_{10}\frac{I_o}{I_e}$ is known as absorbance (A), or optical density (OD). (Please read further comments in Box 9.1.) This value is obtained directly from the instrument (the colorimeter or spectrophotometer). The Beer–Lambert Law only holds under certain conditions (i.e., there are certain limitations), such as the need to use monochromatic light and that it is applicable only at low concentrations, which in a practical sense usually means up to an absorbance of about 0.6.

The unknown concentration of a solution may be determined by one of three approaches:

(a) If the relationship between OD and concentration is not linear, or if it is not established to be linear, then several known concentrations and their ODs are plotted on a graph and the unknown read off the graph (see Figure 9.1).

(b) When it is known that there is a linear relationship between concentration and OD, then for calculation purposes, one point is sufficient. This point (concentration and its corresponding OD) must be higher than the unknowns that are to be calculated. (If indeed one point is used, do not draw one-point graphs. Doing so will be a waste of time and accuracy will be lost in going from numerical values to pictorial representation and back again.) For calculating the concentration of the unknown, a ratio formula, derived as follows, is used:

$$OD_u = Kc_u l$$

$$OD_s = Kc_s l$$

Note that the K and l are the same in both equations. Hence:

$$\frac{OD_u}{OD_s} = \frac{C_u}{C_s}$$

$$C_u = \frac{OD_u}{OD_s} \times C_s$$

There is nothing complicated about this formula; it is almost commonsense (about the ratios, and the formula simply an extension of that). Remember, **this ratio calculation formula** may only be used if it has been established that there is a linear relationship between the concentrations and the ODs *and* the unknown (whose concentration is to be calculated) is within the range of linearity. (Usually, this means that the unknown reading is *smaller* than a higher reading given by a standard.) Readings beyond the linear range may only be "read off the curve" (see Figure 9.1) and not be calculated by the ratio formula.

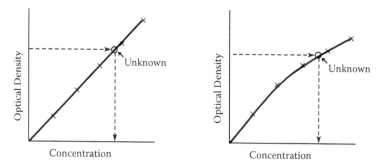

Figure 9.1   Relationship between optical density and concentration.

(c) The third approach is to calculate directly from the Beer–Lambert formula:

$$\text{Absorbance (A)} = Kcl \qquad\qquad (i)$$

It is easy to realise that if concentration (c) and length (l) are both 1, then:

$$\text{Absorbance} = K \ (or \ \varepsilon)$$

In mathematical terms (and keeping in mind that, in chemistry, concentration of 1 means 1 mol/L), this means that K is the absorbance of a *one molar* solution. In practice, this can never be. The absorbance will be ridiculously high and well beyond Beer–Lambert's linearity. However, this K (molar absorbancy or molar absorbancy index or extinction coefficient) is a convenient way to list this value for different substances in data books. For example, the molar absorbance for NADH at 340 nm (A_{340}) is 6.22×10^3. What can usefully be obtained from this information is that 0.1mmol/L NADH would read 0.622. This is because

- 1.0 Molar (mol/L) is supposed to read 6.22×10^3
- 1.0 mmol/L is supposed to read 6.22
- 0.1 mmol/L does read 0.622

Indeed molar absorbancies are obtained by plotting standard curve values in the "readable range" (absorbance 0 to 0.6) and then extrapolating.

The Beer–Lambert formula calculation approach is suitable only when the compounds themselves absorb at particular wavelengths. The molar absorbance is a characteristic constant for that compound such as boiling point, melting point, or specific optical rotation. It is not a variable that is dependent on the method used in producing the particular colour or the absorption peak. In colorimetric methods (e.g., the Somogyi–Nelson method used for measuring reducing sugars or the Folin–Lowry method used for measuring proteins), the exact amount of colour depends on the conditions used in producing the colour. In such cases, there is no meaningful K value and the standard curve or ratio of optical densities approach should be taken.

The Beer–Lambert formula, shown in equation (i), may be rearranged to give:

$$\text{Concentration (mol/L)} = \frac{\text{Absorbance}}{\varepsilon \times \text{pathlength (cm)}}$$

With this transformation, the concentration may be calculated from the absorbance obtained provided molar absorbancy is known or given.

There is one more important point to be realised in calculations involving **colorimetric standard methods.** The word *standard* is best to be taken to mean *the same thing is done to all the tubes.* The only variable is the colour-producing solute (e.g., glucose in the Somogyi–Nelson method). Realise that the final volume is the same for all the tubes. It follows, therefore, that the amount (i.e., absolute amount) of glucose (in mmol, mg, µg, etc.) would be proportional to the OD. Use this value (and not concentration) in calculations and in plotting standard curves. While the Beer–Lambert Law does deal with concentrations (of solutions in the cuvette), the use of concentrations in calculations associated with standard methods often creates confusion. It is proper convention to show amounts (not concentrations) on standard curves (see Figure 12.3).

Box 9.1   Absorbance and Molar Absorbancy Index

Although many authorities allow the terms **absorbance (A)** and **optical density (OD)** to be used interchangeably, it is recommended that OD be used only when dealing *colorimetric* readings (i.e., with wavelengths in the visible range). Absorbance (A) *must be used* for readings from spectrophotometers in *nonvisible* ranges of wavelengths. (Spectrophotometers can be used to read in the visible range as well.) The current trend is to use (only) absorbance (A) for readings at *any* wavelength. OD does not "make sense" in the *nonvisible* ranges of wavelengths. In most biochemistry courses, colorimetry and OD constitute the introductory aspects of this topic; accordingly OD will be mainly used in introductory parts of this chapter.

   Molar absorbancy index (K) is known by various other names, such as **absorbancy index, molar absorptivity** (or just **absorptivity**), and **molar extinction coefficient** (or just **extinction coefficient**). It has also been called **epsilon** — the older symbol for molar absorbancy index. (The term *extinction* crept in here because earlier colorimeters used a process of light extinction with resistance to obtain OD readings; today we have direct reading galvanometers and digital read-out.)

   In relation to chemistry, there are two meanings of the word "absorb" — the fluid, "soak-up," and the other, the photometric or light phenomenon. Generally the "b" extensions of the word, like absorbance and absorbancy, are used for the light phenomenon, and the "p" extensions for the "soak-up" concept. Indeed, there is a "molar absorption coefficient" referring to the "soak-up" phenomenon. Unfortunately, the "b" words are limited — we do not have "absorbation" and "absorbivity," and the "p" words tend to overlap onto both phenomena.

   The term *molar absorbance* is to be avoided because, as explained earlier, it is *not* the absorbance of a 1 molar solution.

   There is also a term **specific absorption index.** In the light phenomenon, it is the absorbance through 1 cm per unit of mass concentration. The unit of mass concentration must be specified when this term is used and can be in gravimetric units, such as g/mL, g/L, or g/100 mL.

Protein and Nucleic Acids Analyses by UV Spectrophotometry

Colorimetric analysis mainly means estimation of colourless substances by causing them to form a coloured complex quantitatively; the complex in turn can be measured in the visible wavelengths in a colorimeter. Some substances are naturally coloured; these of course can be read directly. Then there are some that may be colourless but have absorption in the UV region. These, like the naturally coloured molecules, can be read directly — it is with this kind of molecule, as hinted earlier, that the term *molar absorbancy* (and the direct calculation approach) has relevance.

   What characteristics do these molecules possess? With organic molecules, it is the presence of conjugated double-bond systems. The organic molecules of interest to biochemists are certain vitamins, nucleic acids, and proteins. In inorganic chemistry, interests are aqueous solutions of transition elements and their complexes (due to transition of electrons between the two closely spaced d-orbitals).

Nucleic Acid and Protein Estimations

The presence of conjugated double-bond systems in nucleotides means that deoxyribonucleic acid (DNA) and ribonucleic acid (RNA) absorb light in the UV region, maximally at 260 nm. Likewise, the presence of aromatic amino acid residues (tyrosine, tryptophan, and phenylalanine) in proteins is responsible for proteins having an absorption maximum at 280 nm. There are sensitive quantitative methods based on this phenomenon for measuring both nucleic acids and proteins. However, there are certain problems:

1. The closeness of the absorption peaks means the likelihood of interferences. If either the nucleic acid or the protein is contaminated by the other, the interference in absorption would be considerable.
2. There are variations in the exact absorption peak for the individual amino acids, and because amino acid composition varies from protein to protein, the wavelength for the maxima as well as the value of the molar absorbancy can vary.
3. To some extent this is also true also for the nucleic acids, the 260 peak is a cumulative peak. The absorbance can vary depending on the contributing nucleotides, whether appearing as individual nucleotides, the extent of linkage, the extent of hydrogen bonding between the polynucleotides (i.e., extent of double helical structure), and whether DNA or RNA.
4. Nevertheless, these methods are popular because (a) they are easy to perform (simply read in a spectrophotometer) and (b) they are highly sensitive (only very small amounts are needed for assay). These methods are particularly favoured in monitoring protein or nucleic acids enrichment during separation techniques.
5. In the Warburg-Christian method of protein estimation (originally based on the protein enolase in the presence of yeast nucleic acid), the absorbances at 260 nm and 280 nm are recorded and a correction factor (see Table 9.1), which varies with the ratio of the two readings, is applied to work out the protein concentration. The major points are:
   - As can be seen in Table 9.1, the greater the A_{280} compared to the A_{260}, the less the protein is contaminated by nucleic acid. A correction factor is incorporated in a formula thus:

$$(A_{280}) \text{ (correction factor)} = \text{mg/mL protein.}$$

This formula is applicable for protein concentrations up to an A_{280} of 1.0 in a 1 cm wide cuvette (see Question 9.19).

Table 9.1 Correction Factors used in Warburg-Christian Protein Estimation Method

A_{280}/A_{260}	Correction Factor	Nucleic Acid (%)
1.75	1.12	0
1.36	0.99	1.00
1.16	0.90	2.00
1.03	0.81	3.00
0.87	0.68	5.00
0.75	0.55	8.00
0.67	0.42	12.00
0.60	0.29	20.00

- In nucleic acid estimations, somewhat the converse approach is taken. Pure DNA preparations have an A_{260}/A_{280} of 1.8. Ratios less than this generally indicate contamination by proteins, and ratios higher than 1.8 indicate the presence of RNA, and so forth. An empirical value of A_{260} of 1.000 has been established for 50 μg/mL double-stranded DNA (free from protein contamination) in a 1 cm wide cuvette.

Note: Molecular biology is the new emerging branch of biochemistry. The study of nucleic acids, proteins, and their interaction in protein biosynthesis is a central part of molecular biology. There are important applications of UV spectrophotometry in relation to protein and nucleic acids analyses. The quantitative aspects of this topic are covered in more depth in Chapter 13.

Flame Photometry and Atomic Absorption

Both flame photometry and atomic absorption are suitable for measuring concentration of metal ions in solution. In flame photometry, the metal ion solution is nebulised in a flame, some of the ions move into a higher energy state, and the electrons involved on falling back to the ground state emit light of a particular wavelength. The intensity of this light (i.e., transmission) is measured. A known standard is arbitrarily set at 100%, and lower standards and unknowns are in linear proportion to this standard. With atomic absorption, the light source used is exactly the element under investigation. It emits radiation at precisely the same wavelength as that which will be absorbed by nonexcited atoms of the same element (when their solutions are nebulised and fed into a flame). Being absorption, the Beer–Lambert principles apply. Absorbance values of about 0.6 are the usual maximum for linearity and lower unknown concentrations are worked out in the same manner as in spectrophotometry and colorimetry.

Questions

1. 1.0 mL of a glucose solution (of unknown concentration) was assayed by a standard colorimetric method along with 1.0 mL of a 200 μg/mL standard glucose.

$$OD_{unknown} = 0.2 \quad OD_{std} = 0.2$$

Calculate the concentration of the unknown.

2. 1.0 mL of a glucose solution (of unknown concentration) was assayed by a standard colorimetric method along with 1.0 mL of a 200 μg/mL standard glucose.

$$OD_{unknown} = 0.1 \quad OD_{std} = 0.2$$

Calculate the concentration of the unknown.

3. 1.0 mL of a glucose solution (of unknown concentration) was assayed by a standard colorimetric method along with 0.5 mL of a 200 μg/mL standard glucose.

$$OD_{unknown} = 0.2 \quad OD_{std} = 0.2$$

Calculate the concentration of the unknown.

4. 0.1 mL of a glucose solution (of unknown concentration) was assayed by a standard colorimetric method along with 0.5 mL of a 200 µg/mL standard glucose.

$$OD_{unknown} = 0.2 \ OD_{std} = 0.2$$

Calculate the concentration of the unknown.

5. 1.0 mL of a glucose solution (of unknown concentration) was assayed by a standard colorimetric method along with 0.5 mL of a 200 µg/mL standard glucose.

$$OD_{unknown} = 0.1 \ OD_{std} = 0.2$$

Calculate the concentration of the unknown.

6. It would not be correct to ask a question like the ones above if:

$$OD_{unknown} = 0.8 \ OD_{std} = 0.2$$

Why not?

7. In a laboratory setting, the situation referred to in Question 9.6 actually happens. You get an $OD_{unknown} = 0.8 \ OD_{std} = 0.2$. You decide the correct thing to do is to dilute the UNKNOWN 1 in 5 and repeat the analysis along with the same STD as before. You again get 0.2 for the standard. What reading would you expect for the diluted UNKNOWN? Suggest a range and comment.

8. In the Somogyi–Nelson colorimetric method, 1 mL 200 µg/mL glucose gave an absorbance of 0.38. What absorbance would you expect for 1 mL 100 µg/mL xylose under identical conditions?

$$M_r \text{ glucose } 180 \ A_r \ C \ 12 \ O \ 16 \ H \ 1$$

9. A glucose solution of unknown concentration was diluted 1 in 500 and 1 mL assayed by the Somogyi–Nelson method. The standard (1 mL of 1 mmol/L) gave an optical density reading of 0.51; the unknown 0.17. What is the concentration of the original glucose solution? Give the answer in:
(a) mmol/L.
(b) g /L.

$$M_r \text{ glucose } 180$$

10. Glucose and an unknown sugar-like substance containing C H O only and one oxidisable group were assayed by the Somogyi–Nelson method. Each at identical concentrations (100 µg/mL) gave optical density readings of 0.40 and 0.35, respectively. Calculate M_r of the unknown.

$$M_r \text{ glucose } 180 \ A_r \ C \ 12 \ O \ 16 \ H \ 1$$

11. 3.0 g of glycoprotein, known to contain glucose and mannose, was hydrolysed. The hydrolysate was diluted to 100 mL. 1.0 mL was assayed by Somogyi–Nelson and glucose oxidase methods. The optical densities, along with that of 1.0 mL 100 µg/mL glucose, are recorded in Table 9.2.
(a) Calculate the amount of mannose and glucose in the 100 mL hydrolysate.
(b) Express each as a percentage as they exist in the glycoprotein.

Table 9.2   Data for Question 9.11

	Somogyi–Nelson	Glucose Oxidase
Sample	0.36	0.13
Glucose standard	0.48	0.26

Note: The Somogyi–Nelson method estimates total reducing sugars, and the glucose oxidase method measures glucose only.

12. A mixture known to contain galactose, glucose, and fructose was assayed by different colorimetric methods, as detailed in Table 9.3. Calculate the concentrations of each of galactose, glucose, and fructose, giving your answers in mmol/L.

$$M_r \text{ glucose} = M_r \text{ fructose} = M_r \text{ galactose} = 180$$

13. 1.33 mmol/L NADH is diluted 1:15 and read in a spectrophotometer at 340 nm. What would be the absorbance?

$$\varepsilon_{NADH} \ 6.22 \times 10^3$$

14. A spectrophotometric NADH-340 nm system was used with malate dehydrogenase to estimate the malate concentration in a given solution.

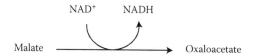

An effective absorbance difference of 0.28 was obtained. (This value is the reading after incubation minus the reading before, and suitably corrected for blanks.) Assuming that the reaction as written moves only 80% to the right, what is the concentration of malate in the assay?

$$\varepsilon_{NADH} \ 6.22 \times 10^3$$

Table 9.3   Data for Question 9.12

	Somogyi–Nelson Method	Roe's Method	Glucose Oxidase Method
Volume of standard glucose or fructose (100 mg/mL)	1.0 mL	1.0 mL	2.0 mL
OD of standard at appropriate wavelength	0.48	0.50	0.40
Volume of mixture used in assays	1.0 mL	0.5 mL	2.0 mL
OD of mixture	0.44	0.125	0.10

Note: The Somogyi–Nelson method estimates total reducing sugars, the Roe's methyl furfuraldehyde method fructose only, and the glucose oxidase method glucose only.

15. You are provided with a freeze-dried extract that is approximately 25 mg NADH/g extract. Calculate the mass of extract you would need to dissolve in 100 mL water in order to obtain an A_{340} of about 0.3.

$$\varepsilon_{NADH}\ 6.22 \times 10^3\ M_r\ 665$$

16. You are provided with a biological extract containing, among other organic acids, lactic acid. The lactic acid concentration is in the order 10–40 mmol/L. A spectrophotometric assay system for measuring, at 340 nm, the NADH produced is available.

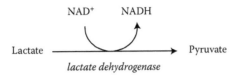

This assay method requires the addition of 0.2 mL of the aliquot (containing the lactate) to 2.8 mL reagents (enzyme, buffer, etc.). What dilution of the given biological extract would you make in order that your dilution fits the requirements of the assay and gives an absorbance between 0.1 and 0.6. A sensible round number dilution is expected.

$$\varepsilon_{NADH}\ 6.22 \times 10^3$$

17. A 1:1000 dilution of serum was read in a flame photometer using a yellow (sodium) filter; the reading was 32. The photometer was standardised using distilled water and 25 μg/mL NaCl; the readings were 0 and 100, respectively. Calculate sodium concentration in serum, give your answer as:
(a) mg sodium/100 mL serum.
(b) mol /L.

$$A_r\ Na\ 23\ Cl\ 35$$

18. Assume that calcium present in blood is bound to a globulin plasma protein. This fraction was separated as described below and analysed for calcium in a flame photometer.
2.0 mL saturated ammonium sulphate was added to 2.0 mL whole serum. The precipitated fraction was dissolved in 250 mL water. This solution was read in a flame photometer using a suitable filter; the reading was 23. The photometer was standardised using distilled water and 0.5 mmol/L $CaCl_2$; the readings were 0 and 100, respectively. Calculate the amount of calcium in serum giving the answer in mol L^{-1} and in scientific notation.

$$A_r\ Ca\ 40\ Cl\ 35$$

19. In the Warburg-Christian protein estimation method, the following absorbencies were recorded:

$$A_{280} = 0.426$$

$$A_{260} = 0.413$$

Calculate protein concentration in g/100 mL.

20. (a) Using zinc sulfate heptahydrate (ZSH), how would you prepare 1 L of 1.00 ppm solution of zinc for use as a standard in atomic absorption?

$$M_r \, ZnSO_4.7H_2O \; 287.5 \; A_r \, Zn \; 65.4$$

(b) Zinc is used as a treatment for footrot disease in sheep. In an experiment, the zinc content in hoof material from sheep that were treated with zinc (T) and controls (C), which were untreated, was analysed. 0.106 g (T) and 0.095 g (C) were each digested in 1.0 mL boiling HCl and then diluted to 100 mL and 20 mL deionised water, respectively. These dilutions, along with the 1.00 ppm zinc standard, were read in an atomic absorption analyser at 214 nm. Readings were C 0.050, T 0.164, Std 0.310. Calculate the zinc content in the tissues. Give answers in µg zinc/g hoof material.

Answers

1. The answer is **200 µg/mL.** There is no need to show working!

2. By inspection, the amount should be **100 µg/mL.** The calculation would be:

$$\frac{0.1}{0.2} \times 200 = 100 \; \mu g/mL$$

3. $\dfrac{0.2}{0.2} \times 100 = 100 \; \mu g$

↑ as 0.5 mL is taken, there is only 100 µg in the standard.
This multiplicand is an absolute amount (100 µg) — not a concentration unit.
Then take the next step:
This 100 µg is contained in this 1.0 mL (taken of the unknown for analysis).
The answer, therefore, becomes **100 µg/mL.**
It is essential that this kind of reasoning be adopted (see Question 9.4).

4. See answer for Question 9.3.
The 100 µg (obtained above) is contained in 0.1 mL of unknown.
Therefore, there is 1000 µg or 1 mg in 1.0 mL unknown.
Answer: 1 mg/mL

5. $\dfrac{0.1}{0.2} \times 100 = $ **50 µg/mL**

6. As the unknown OD is much greater than the standard OD, there may not be a linear response to this (0.8) value (i.e., Beer–Lambert relationship may not hold).

7. See answer to Question 9.6. Diluting 1 in 5, we could expect the reading to be one-fifth of 0.8 or 0.16, but the 0.8 is a "squashed down" reading (usually this is what happens to high — meaning "beyond linear response" — concentrations on a standard curve). The 0.8 is more likely to be higher, consequently, the 0.16 would also be higher. So the expected range is a reading greater than 0.16, possibly 0.2 (if it turns out greater than 0.2, strictly we will have to do it again!). Now that you know this, you may decide to do a 1 in 8 in the first place. Do not however get it too low, because there is always the worry that "the sensitivity will suffer against the background noise."

8. M_r glucose 180, M_r xylose 150 (xylose is a pentose [see Question 4.5]). Both glucose and xylose contain one aldehyde group per molecule. Because xylose is a small molecule, there would be more xylose molecules in 200 µg, and it will consequently have a greater *reducing power* (when the same gravimetric masses are compared). The answer should be:

$$0.38 \times \frac{180}{150} = 0.456 \text{ (if 200 µg)}$$

as only 100 µg considered. OD would be:

$$\frac{1}{2}(0.456) = \mathbf{0.228}$$

Note: The tedious way to go about this would be to find the number of moles of each sugar in 200 and 100 µg. There is really no need.

9. (a) $\dfrac{0.17}{0.51} \times 1$ mol in 1.0 mL diluted unknown.[*]

$$= \frac{0.17}{0.51} \times 1 \times 500 = \mathbf{167 \ mol/L} \text{ undiluted}$$

[*]**Note:** Do include this line — do not directly write "/mL" unless you have thought about it and in your mind said "per mL" and to mean "in 1 mL." It is necessary to be pedantic about this (see Question 9.4).

(b) $\frac{0.17}{0.51} \times 1 \times 500 \times 180 = 30\ 000$ mg/L = **30 g/L**

10. In the 100 µg/mL solutions, the chemical concentration (mol/L) is inversely proportional to relative molecular mass (M_r) and the chemical concentration is directly proportional to the optical density:

$$\text{mol/L} \propto \frac{1}{M_r}$$

where $\propto$ means "is proportional to"
and

$$\text{mol/L} \propto \text{OD}$$

$$\Rightarrow \text{OD} \propto \frac{1}{M_r} \text{ (see Questions 4.5 and 9.8)}$$

$$\Rightarrow M_r = \frac{0.40}{0.35} \times 180$$

$$= \mathbf{206}$$

Note: We expect an M_r greater than 180; only then would there be fewer molecules in the 100 µg to give the lower OD. This is the converse of the xylose situation in Question 9.8.

11. Glucose $= \dfrac{0.13}{0.36} \times 100 = 50 \ \mu g/mL$

$$= 5.0 \ mg \ / \ 100 \ mL \ (\text{take more steps, if you like})$$

Reducing sugars (glucose and mannose)

$$= \dfrac{0.36}{0.48} \times 100 = 75 \ \mu g/mL$$

$$= 7.5 \ mg \ / \ 100 \ mL$$

$$\Rightarrow \text{mannose} = 7.5 - 5.0 = 2.5 \ mg/100 \ mL$$

1 mole of hexose (180 g) loses 18 g H_2O during bond formation with other monosaccharides or amino acid residues (e.g., 100 g glucose becomes 93 g glycogen [see Question 12.2]). Allowing for this, 5.0 mg glucose becomes:

$$5.0 \times \dfrac{162}{180} = 4.5 \ mg$$

% of glycoprotein $= \dfrac{4.5}{3000} \times 100 =$ **0.15 %**

and mannose is half of that

$$= \mathbf{0.075 \ \%}$$

12. Total reducing sugars $= \dfrac{0.44}{0.48} \times 100 = 91.6 \ mg/mL$

$$\text{Fructose} = \dfrac{0.125}{0.50} \times 100 \ mg/0.5 \ mL$$

$$= 25 \ mg/0.5 \ mL \ \text{or} \ 50 \ mg/mL$$

$$\text{Glucose} = \dfrac{0.10}{0.40} \times 200 \ mg/2 \ mL$$

$$= 50 \ mg/2 \ mL \ \text{or} \ 25 \ mg/mL$$

$$\text{Galactose} = 91.6 - (50 + 25)$$

$$= 16.6 \ mg/mL$$

Note: 180 mg/mL = 1 mol/L, and to get mmol/L multiply by 1000

$$\text{Fructose} = \dfrac{50}{180} \times 1000 = \mathbf{278 \ mmol/L}$$

$$\text{Glucose} = \dfrac{25}{180} \times 1000 = \mathbf{139 \ mmol/L}$$

$$\text{Galactose} = \dfrac{16.6}{180} \times 1000 = \mathbf{92.3 \ mmol/L}$$

13. 1 mol/L $A_{340} = 6.22 \times 10^3$

$$1 \text{ mmol/L } A_{340} = 6.22$$

$$\frac{1.33}{15} \text{ mmol/L } A_{340} = \frac{1.33}{15} \times \frac{6.22}{1}$$

$$= 0.552$$

14. 1 mol/L malate conversion at 100% would produce 1 mol/L NADH with an A_{340} of 6.22×10^3.

$$\Rightarrow 0.28 \text{ would correspond to } \frac{0.28}{6.22 \times 10^3} \text{ mol/L}$$

Had all the malate been converted — and not just 80% — the value would be greater:

$$\frac{100}{80} \times \frac{0.28}{6.22 \times 10^3} \text{ mol/L}$$

$$= \frac{100}{80} \times \frac{0.28}{6.22} \text{ mmol/L}$$

$$= 0.056 \text{ mmol/L}$$

15. It would be easiest to go for an A_{340} of 0.311 (because this relates to 6.22×10^3 easily).

$$1 \text{ mol/L NADH } A_{340} = 6.22 \times 10^3$$

$$1 \text{ mmol/L NADH } A_{340} = 6.22$$

$$0.1 \text{ mmol/L NADH } A_{340} = 0.622$$

$$\text{and } 0.05 \text{ mmol/L NADH } A_{340} = 0.311$$

(**Note:** we are now at the *readable level* on spectrophotometers and within Beer–Lambert's linearity.)
We need a 0.05 mmol/L solution of NADH.

$$= 0.05 \times 665 \text{ mg/L}$$

$$= \frac{0.05 \times 665}{10} \text{ mg/100 mL } (= 3.325 \text{ mg})$$

But the material provided is not pure NADH; only 25 mg in 1 g is NADH.
25 mg NADH is present in 1.0 g

$$\Rightarrow \frac{0.05 \times 665}{10} \text{ mg NADH is present in } \frac{0.05 \times 665}{10} \times \frac{1}{25} \times 1\text{g}$$

$$= 0.133 \text{ g}$$

Note: 0.133 g is indeed 40-times greater than 3.325 mg — the same ratio as 1 g $: 25$ mg.

16. 1 mole lactate would produce 1 mole NADH, and 0.1 mmol/L NADH gives an A_{340} of 0.622. At the assay level, the NADH produced should be 0.1 mmol/L (same as the lactate producing it). As there is a 15-fold dilution at the assay (0.2 mL in 3.0 mL), the introduced lactate concentration should allow for this. The required concentration of lactate therefore should be $15 \times 0.1 = 1.5$ mmol/L.

The problem now is, how do we dilute 10–40 mmol/L to give 1.5 mmol/L. Let us consider the higher value, 40 mmol/L:

$$\frac{40}{1.5} = 26.7$$

A 1:26.7 dilution at the assay level would give an A_{340} of 0.622; we want a maximum 0.6. Try 1:30.

A **1:30** 40 mmol/L would be 1.33 mmol/L. Also, a **1:30** 10 mmol/L would be 0.33 mmol/L. 15-fold dilutions of these concentrations would produce assay absorbances of 0.552 and 0.138, respectively. (See Question 9.13 and its answer.)

17. (a) 100 given by $25 \times \dfrac{23}{58}$ μg sodium/mL diluted serum*

$$\Rightarrow 32 \text{ given by } \frac{32}{100} \times 25 \times \frac{23}{58} \text{ μg sodium/mL diluted serum}$$

$$= \frac{32}{100} \times 25 \times \frac{23}{58} \times 100 \text{ μg sodium/100 mL diluted serum}$$

$$= \frac{32}{100} \times 25 \times \frac{23}{58} \times 100 \times 10^{-3} \text{ mg sodium/100 mL diluted serum}$$

$$= 0.317 \text{ mg sodium/100 mL diluted serum}$$

$$= \textbf{317 mg sodium/100 mL whole serum}$$

*Note: $\dfrac{23}{58}$ is used in this step because only 23 parts out of 58 is sodium.

(b) 317 mg/100 mL (from part (a)) = 3170 mg/L

$$= \frac{3170}{23} \text{ mmol/L}$$

$$= \frac{3170}{23} \times 10^{-3} \text{ mol/L}$$

$$= \textbf{0.138 mol/L}$$

Or directly, not using part (a):

Standard NaCl, 25 μg/mL = 25 mg/L

$$= \frac{25}{58} \times 10^{-3} \text{ mol/L}$$

$$100 \text{ given by } \frac{25}{58} \times 10^{-3} \text{ mol/L}$$

$$\Rightarrow 32 \text{ given by } \frac{32}{100} \times \frac{25}{58} \times 10^{-3} \text{ mol/L} = 0.000138 \text{ mol/L}$$

For whole serum, **0.138 mol/L**

Note: The term "mol" is used even for sodium ions. (We do not mean sodium chloride.)

18. Ca in standard $= 0.5 \times 10^{-3}$ mol/L. All Ca from the 2 mL serum is now in 250 mL
100 given by 0.5×10^{-3} mol/L.

$$\Rightarrow 23 \text{ given by } \frac{23}{100} \times 0.5 \times 10^{-3} \text{ mol/L diluted serum}$$

$$= \frac{23}{100} \times 0.5 \times 10^{-3} \times \frac{250}{2} \text{ mol/L whole serum}^*$$

$$= 0.0144 \text{ mol/L}$$

(or as some technical-minded people will have it, and the question asks for it)

$$= 1.44 \times 10^{-2} \text{ mol L}^{-1}$$

***Note:** 2 mL whole serum taken so the dilution is only $\frac{250}{2}$, not 250. This working out can also be approached with an extra step.

19. Ratio $= \dfrac{A_{280}}{A_{260}} = \dfrac{0.426}{0.413} = 1.03$

The correction factor corresponding to this ratio is 0.81 (see Table 9.1) using the formula:

$$(A_{280}) \times (\text{correction factor}) = \text{mg/mL}$$

$$0.426 \times 0.81 = 0.35 \text{ mg/mL}$$

Using the given correction factor, the answer must come out as mg/mL. You can then convert to other concentration units:

$$0.35 \text{ mg/mL} = \textbf{35 mg/100 mL}$$

20. (a) 1 ppm $= 1 \mu\text{g/mL}$

$$= 1 \text{ mg/L}$$

Better prepare 100 mg/L and then dilute 1 in 100, which means we need 100 mg Zn.
65.4 mg Zn contained in 287.5 mg ZSH

$$\Rightarrow 100 \text{ mg Zn contained in } \frac{100}{65.4} \times 287.5 \text{ mg}$$

$$= 439.6 \text{ mg}$$

Dissolve 0.4396 g ZSH in 1.00 L distilled or deionised water; then dilute 10 mL in 1.00 L.

(b) **Control**

$$\frac{0.050}{0.310} \times 1.0 \text{ ppm, or } \mu g/mL$$

$$= \frac{0.050}{0.310} \times 1.0 \times 20 \times \frac{1.0}{0.095} \ \mu g/g$$

$$= \textbf{34 } \boldsymbol{\mu}\textbf{g/g} \text{ hoof material}$$

Treated

$$\frac{0.164}{0.310} \times 1.0 \times 100 \times \frac{1.0}{0.106}$$

$$= \textbf{499 } \boldsymbol{\mu}\textbf{g/g} \text{ hoof material}$$

Note: Check out the dilutions. Control went into 20 mL and treated into 100 mL. Excellent clues for evaluating the order-of-magnitude changes for the two answers. (And the readings contributed to an even greater increase.)

10
Lipid Analysis

Modern methods of lipid analysis centre around gas-liquid chromatography (GLC). Solubility is one of the most striking features of lipids; unlike most biological substances, lipids are not water-soluble. They are soluble to varying degrees in different organic solvents and in different mixtures of organic solvents. Accordingly, vaporising lipids and exposing them to the infinite range of partition characteristics provided by the gas (mobile phase) and liquid (immobile phase) of GLC affords an ideal method for separating this non-water-soluble, closely related group of substances.

Analysis of lipids is a complex task. This is understandable as lipids are a complex group. Various carefully executed separation techniques have to be performed to collect the lipid fraction in an intact state (i.e., without degradation or alteration to the molecules). Once this is achieved, the complex mixture is often further separated into the major classes of lipids and then the constituent molecules are often hydrolysed either as a prerequisite or within the next stage — the analytical procedure. The main issue then is the analysis of the fatty acids produced.

In modern research, GLC remains the major technique in the analysis of fatty acids. But bear in mind, like many other biological molecules, especially the substituted or functional group-bearing fatty acids can be extremely heat-liable. This is where high-power liquid chromatography (HPLC) is beginning to play an important role. As HPLC operates at ambient temperatures, the degradation of heat-sensitive groups is avoided.

In both GLC and HPLC, traces are produced where the location of the peaks provides identification when compared to standards, and the area "under the curve" allows quantitation. In most modern instruments, the integration involved in calculating the area under the curve is provided in the printouts.

Fats from different sources vary in their composition, and fats even from a single source are not made up of the same molecules but are mixtures. It is perhaps for this reason that precise chemical parameters, such as molecular masses, number of carbons in the fat molecule, and number of double bonds, are not used to describe a fat. Instead, empirical parameters that reflect on their size or number of double bonds (degree of unsaturation) or other chemical features are used to characterise whether a sample of fat is pure, adulterated, or spoilt. (In recent times, mass spectrometry is a powerful technique that is employed in elucidating many details about the structure of lipids and fatty acids.)

Although the empirical techniques, on which the questions in this chapter are based, are dated, the questions nevertheless provide exercises in stoichiometry of biochemical molecules (i.e., providing clues to the make-up of lipid molecules):

1. A fat or oil from a particular source (e.g., olive oil) is a mixture of similar molecules. Individual molecules may be different from each other with respect the type and position of the ester-linked fatty acids. However, similar fatty acids (or similar lipid molecules) are often present if the lipids are from any pure source. In the case of olive oil, mainly 16 to 20 carbon fatty acids may be present, and each triacylglycerol molecule contains perhaps two to three double bonds on average. It is this sort of information that we are after.

2. An empirical test, saponification number determination, is used to obtain an indication of the length of fatty acids in fats. **Saponification number** is defined as the number of milligrams of KOH needed to saponify 1 g of fat. To **saponify** is to hydrolyse the fatty acids off the glycerol with alkalis such as KOH or NaOH — the hydrolysed fatty acids appear as salts of fatty acids (soaps). Since each molecule of fat, regardless of its size, requires three molecules of KOH to saponify it, the saponification number would indicate the number of fat molecules in the 1 g of fat. The larger the molecules, the fewer there are per gram of fat. If the fat is assumed to be only triacylglycerols, then the saponification number becomes a measure of the average relative molecular mass of the triacylglycerols (see Question 10.4).

3. **Iodine number** measures the degree of unsaturation of a fat. The unsaturation in turn reflects on the number of double bonds and therefore nature of fatty acids present. When iodine (in practice a more reactive form of halogen) is added to a triacylglycerol containing unsaturated fatty acids, it reacts with the double bonds in the molecule, and the iodine number can be calculated from the amount of iodine absorbed.

$$
\begin{array}{ccc}
\overset{\displaystyle H}{\underset{\displaystyle |}{}} \quad \overset{\displaystyle H}{\underset{\displaystyle |}{}} & & \overset{\displaystyle H}{\underset{\displaystyle |}{}} \quad \overset{\displaystyle H}{\underset{\displaystyle |}{}}
\end{array}
$$

$$
-C = C- \ + \ I_2 \ \longrightarrow \ -\underset{\underset{I}{|}}{C} - \underset{\underset{I}{|}}{C}-
$$

One molecule of iodine (I_2) is added across each double bond. **Iodine number** is defined as the number of grams of iodine absorbed by 100 grams of fat. In this analysis, iodine itself is estimated by titration with thiosulfate.

$$
I_2 \ + \ 2Na_2S_2O_3 \ \rightarrow \ 2NaI \ + \ Na_2S_4O_6
$$

4. Other older estimations include **Reichert–Meissel number** (estimating the number of vola-tile or short chain fatty acids) and **acetyl number** (estimating hydroxyl groups present in some lipids). As previously noted, GLC has replaced these involved and often inconclusive methods for gathering information on the chemical composition of fats.

Questions

1. 0.3 of an oil required 30 mL 0.05 mol/L alcoholic KOH for saponification. Calculate the saponification number.
   M_r KOH 56

2. In one approach to the determination of saponification number an *approximate* concentration of 0.5 mol/L KOH is used. An exactly known volume of the KOH is reacted with an exactly known mass of fat, and the amount of KOH used is determined by titration against standard HCl.

Two reaction flasks were set up, one containing no fat and the other with 3.42 g fat. Exactly 50 mL of approximately 0.5 mol/L KOH was added and the reaction allowed to proceed. After an appropriate time, the contents of each flask were titrated against 0.60 mol/L HCl. The blank flask required 47.81 mL and the sample flask 27.31 mL. Calculate the saponification number.

M_r KOH 56

3. 25.00 mL 0.4210 mol/L KOH is added to 1.5763 g of a sample of peanut oil. After saponification is complete, 8.46 mL 0.2732 mol/L H_2SO_4 is needed to neutralise the excess KOH. Calculate the saponification of the oil.

M_r KOH 56 H_2SO_4 98

4. You are given that the saponification number for a fat (containing only triacylglycerols) is 200. Calculate the average M_r of the triacylglycerols present in the fat. Thus, derive a relationship between average M_r and saponification number.

5. Concentrations of iodine solutions are unreliable because the iodine evaporates rapidly. Hence some form of stable iodine or other halogen but yet reactive in reactions with double bonds is used, and the iodine directly or indirectly estimated by titration with thiosulfate. As with saponification number determination (Question 10.2), the iodine treatment is performed on a sample of fat and without (blank), and the amount of iodine in each flask is determined, in this case, by thiosulfate titration.

Suppose 57.00 mL and 35.20 mL of 0.100 mol/L $Na_2S_2O_3$ were required respectively for the blank and sample titrations and the sample mass used was 0.200 g. Calculate the iodine number.

A_r I 127 M_r $Na_2S_2O_3$ 158

6. 0.3 g of an oil required an effective 30 mL 0.1 mol/L $Na_2S_2O_3$ for titration in the determination of iodine number. This value is the difference: (Blank titration volume) − (Sample titration volume).

Calculate the iodine number.

A_r I 127 M_r $Na_2S_2O_3$ 158

7. Consider a triacylglycerol (M_r 857) containing three double bonds. Calculate:
   (a) its iodine number.
   (b) its saponification number.

A_r I 127 M_r KOH 56

8. The analysis of a particular preparation of fat gave the following values:
   • Saponification No. 200
   • Iodine No. 100

   Assuming that the sample contains only triacylglycerols, calculate the number of double bonds per average molecule of triacylglycerol.

A_r I 127 M_r KOH 56

9. A fat, known to be made up of entirely of 3-double-bond or 4-double-bond containing triacylglycerol molecules, is determined to contain 3.1 double bonds per average molecule. Calculate the percentage of each variety of triacylglycerol.

Answers

1. Saponification No. is mg KOH used by 1 g fat.
   mmol KOH $= 30 \times 0.05$
   mg KOH $\quad = 30 \times 0.05 \times 56 = 84$ mg
   0.3 g fat needs 84 mg KOH

   1 g fat needs $\dfrac{1}{0.3} \times 84$

   $\qquad\qquad$ **= 280 (Saponification No.)**

2. The HCl titrates the KOH present in flasks.
   $HCl + KOH \rightarrow KCl + H_2O$
   HCl required to neutralise blank = $\quad$ 47.81 mL
   HCl required to neutralise sample = $\underline{27.31\ mL}$
   Difference $\qquad\qquad\qquad\qquad \overline{20.50\ mL}$
   This difference $= 20.50 \times 0.60$ mmol HCl, which equals 20.50×0.60 mmol KOH.
   M_r KOH $= 56$
   $\Rightarrow$ mg KOH is $20.50 \times 0.60 \times 56$ mg
   3.42 g fat required $20.50 \times 0.60 \times 56$ mg KOH

   1 g fat requires $\dfrac{20.50 \times 0.60 \times 56}{3.42}$ mg KOH

   $\qquad\qquad$ **= 201 (Saponification No.)**

3. Mol $H_2SO_4 = 8.46 \times 0.2732$ mmol
   $\qquad\qquad\quad = 2.3113$ mmol
   Mol of excess KOH $= 2 \times 2.3113 = 4.6226$ mmol[*]
   Original amount of KOH $= 25 \times 0.4210$ mmol
   $\qquad\qquad\qquad\qquad\quad = 10.525$ mmol
   KOH absorbed by fat $= 10.5250 - 4.6226$
   $\qquad\qquad\qquad\qquad = 5.9024$ mmol
   $\qquad\qquad\qquad\qquad = 5.9024 \times 56$ mg
   $\qquad\qquad\qquad\qquad = 330.5$ mg
   For saponification, 1.5763 g needs 330.5 mg KOH.

   $\Rightarrow$ 1 g needs $\dfrac{1}{1.5763} \times 330.5$

   $\qquad\qquad$ **= 210 (Saponification No.)**

   [*]**Note:** 1 mol $H_2SO_4 = 2$ mol KOH because: $2KOH + H_2SO_4 \rightarrow K_2SO_4 + 2H_2O$

4. A saponification number of 200 means 200 mg combines with 1 g fat.
   Or 200 g combines with 1000 g fat.
   3 mol KOH $(3 \times 56$ g) would combine with 1 mol fat.
   200 g combines with 1000 g fat.

$$3 \times 56 \text{ g combines with } \left[\frac{3 \times 56}{200} \times 1000 \right]^* \text{ g fat}$$

$$= 840 \text{ g fat}$$

Remember, we argued that 3×56 g combines with 1 mol of fat, which here worked out as 840 g. Hence, **840 is the M_r.**

*Note: This expression is the relationship between saponification number and molecular weight:

$$\textbf{Average } M_r = \frac{3 \times 56 \times 1000}{\textbf{Saponification No.}}$$

5. Note that the reaction for thiosulfate-iodine titration is:

$$I_2 + 2Na_2S_2O_3 \rightarrow 2NaI + Na_2S_4O_6$$

2 moles of thiosulfate react with 1 mole I_2.

The difference in titre $(57.00 - 35.20)$ is 21.80 mL.

This equals 21.80×0.1 mmol $Na_2S_2O_3$,

which in turn equals $21.80 \times 0.1 \times \frac{1}{2}$ mmol iodine,

which equals $21.80 \times 0.1 \times \frac{1}{2} \times 254$ mg I_2 (as $M_r\, I_2 = 2 \times 127 = 254$).

Iodine no. is g iodine absorbed by 100 g fat. 0.200 g sample absorbs:

$21.80 \times 0.1 \times \frac{1}{2} \times 254$ mg iodine

$\Rightarrow 100$ g sample absorbs $\dfrac{100}{0.2} \times 21.80 \times 0.1 \times \frac{1}{2} \times 254$ mg iodine

$$= \textbf{138 g (Iodine No.)}$$

Note: There are two 1:2 factors involved. The first is one mole of iodine (I_2) reacts with two moles of thiosulfate. ($Na_2S_2O_3$). The other, one molecule of iodine (I_2) is made up of two atoms of iodine. Do not think about this; just apply the right corrections at the right time. (If you leave them both out, you will get the right answer — but you will not be happy with that.)

6. 2 moles $Na_2S_2O_3$ react with 1 mole iodine (see Answer to Question 10.5).

30 mL 0.1 mol/L $Na_2S_2O_3 = 30 \times 0.1$ mmol $Na_2S_2O_3$

$= 30 \times 0.1 \times \frac{1}{2}$ mmol iodine

0.3 g oil absorbs $30 \times 0.1 \times \frac{1}{2} \times 254$ mg iodine

100 g oil absorbs $\dfrac{100}{0.3} \times 30 \times 0.1 \times \frac{1}{2} \times 254$ mg

$$= \textbf{127 g (Iodine No.)}$$

7. (a) Iodine no. = g iodine/100 g fat

1 mole fat (containing three double bonds) combines with 3 moles iodine (i.e., 857 g combines with $3 \times [2 \times 127]$ g iodine = 762 g

$\Rightarrow 100$ g combines with $\dfrac{100}{857} \times 762$ g

$$= \textbf{89 (Iodine No.)}$$

(b) Saponification No. = mg KOH/g fat

There are three fatty acid residues per triacylglycerol.

Each needs 1 mole (56 g) KOH for hydrolysis.

$$\Rightarrow M_r = \frac{3 \times 56 \times 1000}{\text{Sap No.}} \quad \text{(from Answer to Question 10.4)}$$

$$\text{i.e., } 857 = \frac{3 \times 56 \times 1000}{\text{Sap No.}}$$

$$\text{Sap No.} = \frac{3 \times 56 \times 1000}{857}$$

$$= \mathbf{196}$$

8. Saponification No. = mg KOH/g fat; Iodine no. = g iodine/100 g fat

$$\text{Average } M_r \text{ triacylglycerol (TAG)} = \frac{3 \times 56 \times 1000}{\text{Sap No.}}$$

$$= \frac{3 \times 56 \times 1000}{200} = 840$$

100 g fat require 100 g iodine

$$\Rightarrow \frac{100}{M_r \text{ TAG}} \text{ moles fat require } \frac{100}{M_r I_2} \text{ moles iodine}$$

$$\text{i.e., } \frac{100}{840} \text{ moles fat require } \frac{100}{254} \text{ moles iodine}$$

$$\text{and 1 mole fat requires } \frac{840}{100} \times \frac{100}{254} \text{ moles iodine}$$

$$= 3.31$$

Therefore, there are **3.31 double bonds per average molecule.**

Note: Do not round-off the answer to exactly 3. The answer 3.31 says there is an average of 3.31 double bonds per molecule of triacylglycerol.

9. Use 1 = 100%. Let x = 3 db variety and 1 − x = the 4 db variety.

x(3) + (1 − x)(4) = 3.1

3x + 4 − 4x = 3.1

−x = −0.9

x = 0.9, or 90% (and the other 10%)

3 double bonds variety = 90%

4 double bonds variety = 10%

Note: That is why the average is so much closer to the 3 bd variety.

11
Tissue and Fluid Content

Almost all quantitative biochemistry relates to working out something and relating it to an original reference material. In chemistry problems, it is working out number of carbons, number of double bonds, or the like and relating to a molecule (or mole) or a structure; we called this *stoichiometry* in Chapter 4.

In biochemistry, the "bio-" part of the subject also has its quantitative implications. A great deal — if not all — of quantitative (biological) biochemistry involves relating values to a reference weight of a tissue or a body fluid. This is the purpose of the questions in this chapter; admittedly, questions of this nature also appear in other chapters under appropriate techniques or class of biochemical molecules.

Questions

1. 1.86 g of rat liver was digested in 2 mL 30% KOH. The glycogen present in the KOH was precipitated by the addition of 2.5 mL 95% ethanol and centrifuged. The precipitate was dissolved in 100 mL distilled water. 1.0 mL of this solution was hydrolysed with 1.0 mL 1.2 mol/L HCl. After appropriate neutralisation, the volume was adjusted to 5.0 mL with distilled water. 1.0 mL of the diluted hydrolysate was assayed by the Somogyi–Nelson method, along with 1.0 mL of 200 µg/mL standard glucose. The ODs were 0.250 and 0.500, respectively. Calculate grams of glycogen/100 g liver bearing in mind that 1 g glucose is equivalent to 0.93 g glycogen.

2. A protein fraction (A) was precipitated from 2 mL of serum. The more soluble proteins (fraction B) remained in solution. After centrifugation, the supernatant containing fraction B was quantitatively transferred to a calibrated test tube and diluted to 5 mL. Protein present in 1.0 mL of this dilution was treated with trichloroacetic acid (TCA), and the resulting precipitate collected and redissolved in 50 mL of 0.5 mol/L NaOH. 0.05 mL of the 50 mL was assayed by the Folin-Ciocalteu Method, along with 0.5 mL of a 400 µg/mL standard protein.

   OD standard 0.50 OD sample 0.05

   Calculate the amount of fraction B proteins. Give your final answer as g/100 mL serum.

3. Assume that the average body mass of an adult is 70 kg and 80% of body mass is water. Assume too that ingested alcohol is quickly equilibrated throughout the entire body water and none is lost in expired air or urine. What volume of beer (ethanol, 5% w/v) would be required to give a blood alcohol concentration of 0.05% w/v?

4. Perform a one- or two-step conversion of the answer you obtained for the last question to obtain new volumes (that will produce a blood alcohol concentration of 0.05% w/v) for each of the other alcoholic beverages shown in Table 11.1. Do this also for different body weights and enter the entire lot of values in Table 11.1. This question provides a fair amount of data handling exercise (without too much effort) and, if you like playing around on the computer, try the suggestion below. Together with the worked-out answer, you will find advice on how to derive a formula for handling repeated calculations and on effective table presentation.

(You may wish to enter the table on a spreadsheet and even personalise it for your body weight. Try setting up a formula that will generate a whole row of values for each alcoholic beverage as soon as a new body mass is entered.)

5. 0.3 g of K-cells was obtained from 32 g of a tissue. The K-cells were then dissolved in 100 mL of buffer. 2.0 mL of this solution was treated as follows:

A protein fraction (P) was precipitated and redissolved in 4.0 mL water. 1.0 mL of this solution was treated with 1.0 mL TCA and the resulting precipitate was collected by centrifugation and redissolved in 5.0 mL 0.5 mol/L NaOH. 0.5 mL of the 5.0 mL was assayed by a standard colorimetric method along with 1.0 mL of a 100 μg/mL standard protein.

OD standard 0.50 OD sample 0.32

Calculate grams of the protein (P) in

(a) 100 g K-cells.

(b) 100 g tissue.

6. A commercial preparation Krivite contains 36 mg riboflavin/100 g. For an analysis, a solution containing 0.05 μg riboflavin/mL is required. What mass of Krivite would you take, and how would you prepare the required dilution?

7. 3.0 g tissue was dissolved in 100 mL buffer. Calculate amounts of protein, glucose, and other reducing sugars giving each as g per 100 g tissue.

(a) 2.0 mL was treated with 1.0 mL TCA; precipitate formed was centrifuged and dissolved in 5.0 mL 0.5 mol/L NaOH. 0.5 mL of the 5.0 mL was assayed by a standard colorimetric method along with 1.0 mL 200 μg/mL standard protein.

Table 11.1   Blank Table for Question 11.4

Body Mass		Volume of Alcoholic Beverages				
kg	lbs	Light Beer (2.7%)	Regular Beer (5%)	Wine (12%)	Fortified Wine (18%)	Spirits/Hard Liquor (40%)
55	121					
70	154		560 mL			
85	187					
100	220					
115	254					

OD standard 0.50 OD sample 0.32

(b) 0.5 mL and 1.0 mL of the original 100 mL were used in Somogyi–Nelson (S–N) and glucose oxidase (GO) methods, respectively, along with 1.0 mL 200 µg/mL standard glucose in each case.

S-N Method OD standard 0.50 OD sample 0.26

GO Method OD standard 0.58 OD sample 0.32

Answers

1. $\dfrac{0.250}{0.500} \times 200$ µg glucose in 1 mL hydrolysate assayed

$$= \frac{0.250}{0.500} \times 200 \times \frac{5}{1} \text{ µg glucose in 5 mL hydrolysate}$$

$$= \frac{0.250}{0.500} \times 200 \times \frac{5}{1} \times \frac{100}{1} \text{ µg glucose in 100 mL "precipitate solution"}$$

All this came from 1.86 g liver:

$$= \frac{0.250}{0.500} \times 200 \times \frac{5}{1} \times \frac{100}{1} \times \frac{100}{1.86} \text{ µg glucose/100 g liver}$$

$$= \frac{0.250}{0.500} \times 200 \times \frac{5}{1} \times \frac{100}{1} \times \frac{100}{1.86} \times 0.93 \times 10^{-6} \text{ g glycogen/100 g liver}$$

$$= \textbf{2.5 g glycogen/100 g liver}$$

Note: The factor 0.93 makes the answer *smaller* (remember, water is lost when glucose units link up to form glycogen); the 10^{-6} converts µg to g.

Other hints for performing calculations of this type:

a. Start the calculations at the assay level and work up to the tissue level.

b. Do not create imaginary dilution factors. If 1 mL out of 5 mL is assayed, a factor $\frac{5}{.1}$ would allow for calculation of what is in 5 mL. Similarly, if 0.5 mL out of 5 mL is assayed, write $\frac{5}{0.5}$ (not 10), and if 0.5 mL out of 100 mL, write $\frac{100}{0.5}$ (not 200). That is, work with real figures — not what you have worked out in your head (see Chapter 1).

c. Whenever possible, leave expressions as they are. Do not work them out because that would be extra work and introduces the possibility of errors and round-off errors (again, see Chapter 1). If you find the repetitious writing of expressions tedious, do not copy the whole expression from an immediately prior step — simply write down, for example, "(above) $\times \frac{5}{1}$" (for the second line of the calculation) and following that "(above) $\times \frac{100}{1}$." [For reasons of clarity, complete expressions are repeated in this book.]

d. Work with one kind of unit all the way, usually the unit from the assay (µg in this example) and convert that unit to the unit required for the final answer only at the end (to g in this example).

e. Be careful how conversions are done. Do you want the answer bigger or smaller? Do you divide or multiply, and how do you enter the expression? Crazy mistakes can occur even at this point of your calculations. (Please reread the "Multiplication and Division Errors" section in Chapter 2 for more information.)

f. Give a label to each step. This will help you keep track (e.g., hydrolysate, "precipitate," original solution, the dilution, concentrate, etc.). That is, make statements such as "in 5 mL of the hydrolysate...," even if these words do not appear in the question.

g. With experience and confidence, the last line can be written straight away as a *string of blocks*. (Unless you are dealing with similar kinds of problems on a regular basis, it is not recommended you try to do this.)

h. Finally, as always, check that your answers are sensible: 2.5×10^3 g glycogen/100 g tissue has to be a wrong answer.

2. (worked as string of blocks)

$$\overset{A}{} \quad \overset{B}{} \quad \overset{C}{} \quad \overset{D}{}$$

$$\frac{0.05}{0.50} \times 200 \times \frac{50}{0.05} \times \frac{5}{1} \; \mu g/2 \text{ mL serum}$$

$$\overset{E}{} \quad \overset{F}{}$$

$$= \frac{0.05}{0.50} \times 200 \times \frac{50}{0.05} \times \frac{5}{1} \times \frac{100}{2} \times 10^{-6} \text{ g/100 mL}$$

$$= \textbf{5 g/100 mL}$$

The ratios in the calculation are:

A   ratio OD
B   *μg protein in standard (only 0. 5 mL of 400 μg/mL taken)*
C   *only 0. 5 mL of the 50 mL (NaOH) solution assayed*
D   *only 1 mL of the 5 mL (supernatant) solution used*
E   *to convert to 100 mL serum*
F   *to convert μg (the unit used in the calculation because the standard protein is in μg) to g divide by 10^6 (or multiply by 10^{-6})*

3. 70 kg body weight $= 70 \times \dfrac{80}{100}$ L liquid

100 mL to contain 0.05 g alcohol

$$\Rightarrow 70 \times \frac{80}{100} \times 1000 \text{ mL to contain } \frac{70 \times 80 \times 1000 \times 0.05}{100 \times 100} \text{ g alcohol}$$

There is 5 g alcohol in 100 mL beer.

$$\frac{70 \times 80 \times 1000 \times 0.05}{100 \times 100} \text{ g alcohol in } \frac{70 \times 80 \times 1000 \times 0.05 \times 100}{100 \times 100 \times 5} \text{ mL}$$

$$= \textbf{560 mL}$$

4. Look at the last expression in the answer for the last question (reproduced below):

$$\frac{70 \times 80 \times 1000 \times 0.05 \times 100}{100 \times 100 \times 5}$$

The body mass (70) appears in the numerator, and the alcohol concentration of beer (5) appears in the denominator. As you would intuitively know, with higher body mass, higher amounts can be consumed before reaching 0.05% blood alcohol level; conversely, with higher alcohol concentration, a lower volume would be required to reach the same blood alcohol level. Consequently, when an 85 kg body mass drinks the same 5% beer as the 70 kg body mass, the 560 mL calculated in the last question becomes:

$$560 \times \frac{85}{70} = \textbf{680 mL} \tag{i}$$

Whereas, if the alcohol concentration is changed to wine (12% alcohol) but the body mass remains the same 70 kg, the calculation would be:

$$560 \times \frac{5}{12} = \textbf{233 mL} \tag{ii}$$

If both body mass *and* alcohol concentration are changed, as with an 85 kg body weight drinking 12% wine, the calculation becomes:

$$560 \times \frac{85}{70} \times \frac{5}{12} = \textbf{283 mL} \tag{iii}$$

If you wish, you can rearrange the above equation (iii) to

$$560 \times \frac{5}{70} \times \frac{85}{12}$$

and create a single factor for $560 \times \dfrac{5}{70}$

That single factor is 40: $560 \times \dfrac{5}{70} = 40$

Our "conversion formula" then becomes:

$$40 \times \frac{\text{Body Mass}}{\text{Alcohol Concentration}}$$

Comments:

- This idea of combining factors and creating one value that is then used to multiply the variables is not a bad idea at all, especially when you have to perform a large number of identical calculations. (Table 11.2 was completed using this "conversion formula method.")
- With the answers tabulated, as we have here, we have an built-in setup for evaluating our answers — looking for patterns, uniform increments, or trends (also seen earlier in Table 5.5). To achieve this, tables, such as graphs, must be presented in the most effective manner: simply have the values *increasing* from left to right and from top to bottom. *Always try to present tables in this manner.*

Table 11.2   Answer Table for Question 11.4

Body Mass		Volume of Alcoholic Beverages				
kg	lbs	Light Beer (2.7%)	Regular Beer (5%)	Wine (12%)	Fortified Wine (18%)	Spirits/Hard Liquor (40%)
55	121	815 mL	440 mL	183 mL	122 mL	55 mL
70	154	1037 mL	**560 mL**	233 mL	156 mL	70 mL
85	187	1259 mL	680 mL	283 mL	189 mL	85 mL
100	220	1481 mL	800 mL	333 mL	222 mL	100 mL
115	254	1704 mL	920 mL	383 mL	256 mL	115 mL

- The volumes for the "average" 70-kg adult shown in the table are approximately equal to two the "standard drinks" recommended by Australian authorities as the safe amount that may be consumed in 1 hour before reaching the 0.05% blood alcohol concentration (BAC). *[There are legal implications relating to BAC limits. The limits themselves and relevant laws vary from country to country and from state to state.]*

5. (a) $\dfrac{0.32}{0.50} \times 100 \times \dfrac{5}{0.5} \times \dfrac{4}{1} \times \dfrac{100}{2} \times \dfrac{100}{0.3} \times 10^{-6}$ g/100 g K cells

   = 42 g/100 g K cells

   (b) $\dfrac{0.32}{0.50} \times 100 \times \dfrac{5}{0.5} \times \dfrac{4}{1} \times \dfrac{100}{2} \times \dfrac{100}{32} \times 10^{-6}$ g/100 g K cells

   = 0.4 g/100 g tissue

Notes:

- Answer (b) does not involve the figure 0.3 g of K-cells. The question could just as well be: 32 g of tissue was dissolved in 100 mL of buffer....
- There is 0.3 g K-cells per 32 g tissue, or approximately $\frac{1}{100}$th the mass of tissue. Hence, answer (b) should also be approximately $\frac{1}{100}$th that of (a). (As has been maintained throughout this book, evaluate your answers any way you can.)

6. What is this question all about? It appears very likely that Krivite is a paste. We cannot directly weigh out the riboflavin, so we have to weigh out the Krivite. We must take a larger mass of paste, knowing that 100 g of Krivite contains 36 mg riboflavin. Riboflavin wanted is 0.05 µg for 1 mL or 0.05 mg for **1000 mL**. For 0.05 mg we need to take:

$$\dfrac{0.05}{36} \times 100 \text{ g}$$

$$= 0.1389 \text{ g}$$

Masses of this magnitude can easily be weighed in a chemical balance. "Approximately but exactly" weigh out a mass of Krivite just under 0.1389 g. Dissolve this in a little water and dilute to the required mark in a 1000 mL measuring cylinder (or just under half the mass in 500 mL cylinder).

Comments: If you need to, please look up the term "Approximately but exactly" in Chapter 2.

7. (a) **Protein**

$$\frac{0.32}{0.50} \times 200 \times \frac{5.0}{0.5} \times \frac{100}{2} \times \frac{100}{3} \times 10^{-6} \text{ g/100 g}$$

= 2.13 g/100 g

(b) **Reducing sugars**

$$\frac{0.26}{0.50} \times 200 \times \frac{100}{0.5} \times \frac{100}{3} \times 10^{-6} \text{ g/100 g}$$

= 0.693 g/100 g

Glucose

$$\frac{0.32}{0.58} \times 200 \times \frac{100}{1} \times \frac{100}{3} \times 10^{-6} \text{ g/100 g}$$

= 0.368 g/100 g

Other reducing sugars

$$0.693 - 0.368 = \textbf{0.325 g/100 g}$$

12
Practical Calculations

Courses in biochemistry generally include a practical component, and most of the experiments in the practical component involve (biochemical) calculations. The experiments in the practical program in different courses can vary considerably, and so too the necessary calculations. The questions included in this chapter relate to the author's teaching experience. The questions are written in a stand-alone fashion, meaning that all the necessary information needed for solving the questions is provided in the questions themselves. Students are invited to attempt them for two reasons: (a) the questions directly or indirectly will assist in problems related to their practical programs and (b) these questions represent a real biochemical experimental approach of investigation: working out values or discovering principles (undergraduate students, usually, *illustrate* principles). The biochemical techniques relating to the problems (questions) in this chapter have all been introduced in the earlier chapters.

There is one learning exercise that students could benefit from while doing or appreciating these questions. That learning exercise relates to your practical work. Invariably, your lecturers would want you to read your practical notes *before* actually doing the experiments—and your lecturers would have discouraged you from doing them "recipe-book" style. During the prereading, usually towards the end, when it came to working out something, you would no doubt ask yourself: "I will need a certain piece of information. Now where (in the experimental procedure), do I get that information?" The very thought that goes through your mind, "Where do I get that information from," will benefit you enormously in understanding the experiment and performing it intelligently and purposefully during the laboratory session. That thought process "where do I get that information" or "I need such and such a piece of information," whether in relation to practical work or in attempting biochemical calculations, means you are taking control, you are gaining understanding, you are exercising authority over the quantitative logic of biochemistry. That, exactly, has always been the purpose of this book.

Questions

1. 1.0 g heart muscle was homogenised in a final volume of 100 mL phosphate buffer and used as the enzyme source in an experiment designed to study competitive inhibition of succinic dehydrogenase by malonate. Methylene blue was used as the indicator; methylene blue decolourises (from blue to colourless) under the anaerobic conditions used in the experiment.

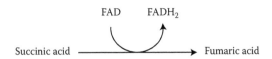

FAD is the prosthetic group of the enzyme succinic dehydrogenase and is reduced to FADH2, which in turn reduces the methylene blue indicator. Assume that the mole relationship between succinic acid, $FADH_2$, and methylene blue is 1:1:1. The tubes were incubated at $37°$ and the time taken for decolourisation is shown in Table 12.1.

(a) Calculate the *amounts* of succinate and malonate present in each of the tubes.

(b) Calculate the final *concentration* of succinate and malonate present in each of the tubes.

(c) In tube No. 1, did the decolourisation occur because all the succinate was used up or because all the methylene blue was used up? Given the M_r of methylene blue is 319, calculate the amount of methylene blue in tube No. 1 and thus compare the amounts of methylene blue and succinate present.

(d) Calculate the number of units of the enzyme succinic dehydrogenase present in tube No. 1.

(e) Calculate units (and katals) of succinic dehydrogenase in 1 g of heart muscle.

2. Two rats were used in an experiment to determine the effect of fasting. Rat A was fasted for 72 hours and Rat B was fed normally. The rats were killed and samples of liver and muscle were removed from each. The tissues were dissolved in 2 mL of 30% KOH. Glycogen present in the KOH digest was precipitated by the addition of 2.5 mL 95% ethanol and centrifuged. The supernatant was collected in labeled test-tubes and the pH adjusted to 7.0 and volume to 10.0 mL.

The precipitated glycogen was dissolved in the volumes shown in Table 12.2. 1.0 mL from each of the solutions was hydrolysed with 1.5 mol/L HCl at $100°$ for 2 hours. After hydrolysis, the pH was adjusted to 7.0 and volume to 10.0 mL.

Glucose present in 1 mL of the 10 mL hydrolysate (or supernatant) was assayed by the Somogyi–Nelson method and the OD readings recorded (see Table 12.2). OD 1 mL standard glucose 100 mg/mL was 0.24.

(a) Calculate the glycogen present in each tissue; give your answers in grams per 100 g tissue. Use the relationship that 0.93 g glycogen when hydrolysed produces 1.0 g glucose.

(b) What information can be obtained from the OD readings of the supernatant? Calculate this and again express the answer in grams per 100 g tissue.

(c) Tabulate your answers.

Table 12.1 Data for Question 12.1			
Tube No.	1	2	3
0.01 % methylene blue (mL)	0.5	0.5	0.5
0.1 mol/L Na succinate (mL)	0.5	0.5	1.0
5 mmol/L Na malonate (mL)	–	0.5	0.5
H_2O (mL)	1.0	0.5	–
Heart muscle homogenate (mL)	1.0	1.0	1.0
Time of decolourisation (min)	5	10	7

Table 12.2   Data for Question 12.2

	Fed		Fasted	
	Liver	**Muscle**	**Liver**	**Muscle**
Weight (g)	1.5	1.5	1.5	1.5
Volume in which glycogen was dissolved (mL)	100	10	10	10
OD hydrolysed sample	0.20	0.22	0.17	0.17
OD supernatant	0.06	0.06	0.05	0.08

OD 1.0 mL 1 μg/mL Standard glucose 0.24

3. An experiment was performed to compare the rates at which glycogen was depleted under anaerobic conditions in rat liver and mussel mantle tissue. Weighed tissues were incubated at 25° for 30 minutes and 1 hour.

   Glycogen Estimation: The tissues were dissolved in 2 mL of 30% KOH. Glycogen present in the KOH digest was precipitated by the addition of 2.5 mL 95% ethanol and centrifuged. The precipitated glycogen was dissolved in either 50 mL or 150 mL water, as shown in Table 12.3. 1.0 mL of either volume was neutralised and the volume adjusted to 10 mL.

   Glucose present in 0.5 mL of the 10 mL hydrolysate was assayed by the Somogyi–Nelson method along with 0.5 mL of standard glucose (200 μg/mL). The OD readings obtained are included in Table 12.3.
   Calculate, remembering that 100 g glucose = 93 g glycogen, the grams of glycogen per 100 g of tissue, and plot the values on a graph.

4. Assume that the equilibrium constant for the following transaminase reaction is 1.5 at 38°.

$$\text{Glutamate} + \text{Pyruvate} \rightleftharpoons \alpha\text{-Ketoglutarate} + \text{Alanine}$$

   (**Note:** The associated experimental details are not relevant for this question.)
   (a) Calculate the standard free energy change.
   (b) Calculate the amount of pyruvate that would be present at equilibrium when initial concentrations of glutamate, α-ketoglutarate, and alanine were each 5 mmol/L (and pyruvate concentration was zero).

Table 12.3   Data for Question 12.3

	Zero Time		30 Minute		1 Hour	
	Mantle	**Liver**	**Mantle**	**Liver**	**Mantle**	**Liver**
Weight (g)	1.3	1.4	1.4	1.5	1.2	1.6
Volume (mL)	150	150	50	50	150	50
OD	0.37	0.32	0.15	0.15	0.30	0.06

OD Standard 0.40

Table 12.4	Volume and Protein Concentration of Fractions	
Fraction	Volume (mL)	Protein Concentration (mg/mL)
Homogenate	40	3.62
Nuclear	20	1.62
Mitochondrial	20	1.58
Microsomal	20	1.51
Final supernatant	40	0.82

Given:

$R = 8.3145$ J K^{-1} mol^{-1}
$0°C = 273$ K
$\ln x = 2.303 \log x$

5. A tissue was homogenised in 50 mL of citrate buffer; 40 mL of this homogenate was differentially centrifuged to obtain the nuclear, mitochondrial, and microsomal fractions, which were then resuspended in 20 mL of citrate buffer. The protein concentration of the original homogenate, the three particular fractions, and the final supernatant was determined and is shown in Table 12.4. The 10 mL of the remaining original homogenate was diluted to 40 mL. The fractions were assayed for glucose-6-phosphatase (G6Pase) activity:

$$\text{glucose-6-phosphate} \rightarrow \text{glucose} + P_i$$

G6Pase activity of each fraction was assayed as follows: 0.5 mL aliquots were incubated with 1.5 mL 2.0 mmol/L glucose-6-phosphate at 25° for 10 minutes. Reactions were stopped by the addition of 8.0 mL of 5% trichloroacetic acid. Zero time controls were prepared by the addition of trichloroacetic acid before the addition of glucose-6-phosphate. Nonenzymatic blanks were set up by incubating 0.5 mL citrate buffer with 1.5 mL 2.0 mmol/L glucose-6-phosphate, followed by the addition of 8.0 mL trichloroacetic acid. All tubes were centrifuged for 5 minutes, and 5 mL of the supernatant was analysed for inorganic phosphate by the colorimetric method of Fiske and Subbarow (Table 12.5).

Table 12.5	Absorbance of Samples	
Fraction	Zero Time	10 minutes
Homogenate	0.16	0.56
Nuclear	0.14	0.32
Mitochondrial	0.15	0.26
Microsomal	0.16	0.69
Final supernatant	0.18	0.24
No enzyme	0.02	0.04

Table 12.6   Absorbance of Inorganic Phosphate

Tube No.	Phosphate Solution (mL)	Water (mL)	Optical Density
1	0.0	5.0	0.00
2	1.0	4.0	0.21
3	2.0	3.0	0.42
4	3.0	2.0	0.60
5	4.0	1.0	0.80
6	5.0	0.0	0.97

A standard curve was prepared for phosphate absorbance using varying volumes of 0.4 mmol/L inorganic phosphate (Table 12.6).

(a) Using values from Table 12.6, plot the standard curve for inorganic phosphate. Using values off the standard curve, calculate the number of units of G6Pase per millilitre for each fraction.

(b) Calculate the specific activity of G6Pase (in units/mg protein) for each fraction.

(c) Calculate the degree of enrichment in the specific activity of the enzyme for each fraction relative to the original homogenate (express this value as a percentage, taking the value of the homogenate as 100%).

6. Deoxyribonucleic acid (DNA) extraction from tissues, as would be expected, is a complicated process. In this question, the details of the extraction process are left out and only the quantitative aspects considered.

The DNA of the white cells from 10 mL whole blood was extracted and resuspended in 200 μL buffer. 10 μL of this solution was mixed with 990 μL distilled water and the absorbance determined at 260 and 280 nm as 0.302 and 0.179, respectively.

(a) Given that 50 μg/mL double-stranded DNA gives an A_{260} of 1.000 and assuming that the preparation is 100% pure and double-stranded, calculate the DNA concentration, giving your answer as milligrams per millilitre (μg/mL) whole blood.

(b) Calculate the OD_{260}/OD_{280} ratio. Given that this ratio is 1.8 for pure double-stranded DNA and that there is a linear fall off of this value as DNA is contaminated with protein, calculate the percentage purity of the preparation.

Answers

1. (a) Amounts of succinate, 0.1 mol/L = 0.1 mmol/mL = 100 μmol/mL

Tube No. 3 contains 100 μmol.

Tube No. 1 and 2 contain 50 μmol each.

Amounts of malonate, 5 mmol/L = 5 μmol/mL

Tube No. 2 and 3 contain 2.5 μmol each.

(b) Final volume in each tube is 3 mL. Final concentrations are:

Succinate

Tube No. 3 100 µmol/3 mL

= 33.3 µmol/mL

= 33.3 mmol/L

(or, can be worked out as $1/3 \times 0.1$ mol/L added = 0.0333 mol/L, which equals 33.3 mmol/L)

Tube No. 1 and 2   **16.65 mmol/L each**

Malonate

Tube No. 2 and 3 are $1/6 \times 5$ mmol/L = **0.83 mmol/L**

Evaluate: In Tube No. 2 [succinate]:[malonate]

= 100 mmol/L:5 mmol/L

= 20:1

Answers 16.65 mmol/L:0.83 mmol/L also equals 20:1

(c) This experiment was designed such that the methylene blue is "used up." In other words, a fast reaction produces a certain amount of $FADH_2$ that decolourises all of the methylene blue present within a certain time. A slow reaction, on the other hand, proceeds for a longer time because it takes longer to produce the same amount of $FADH_2$ to decolourise the same amount of methylene blue. Therefore, we would expect less methylene blue than succinate (of course, the "less" is taken in a moles-sense and not gravimetric-sense).

Moles of methylene blue are calculated for the 0.5 mL of 0.01%:

$$0.01 \text{ g}/100 \text{ mL} = \frac{1}{2} \times \frac{0.01}{100} \text{ g}/0.5 \text{ mL}$$

$$= \frac{1}{2} \times \frac{0.01}{100} \times \frac{1}{319} \times 10^6 \text{ µmoles}/0.5 \text{ mL}$$

$$= \textbf{0.157 µmol/0.5 mL}$$

As expected, methylene blue (0.157 µmol) is much less than succinate (50 µmol). To answer the question: **All the methylene blue was used up.**

(d) The 0.157 µmol of methylene blue = $FADH_2$

= succinate converted in 5 minutes.

1 unit = 1 µmol S converted/minute

0.157 µmol converted in 5 minutes

In 1 minute, $\dfrac{0.157}{5}$ = 0.0314 µmol

= 0.0314 units

(e) 0.0314 units present in 1 mL homogenate

$$= 0.0314 \times 100 = 3.14/100 \text{ mL}$$

$$= \textbf{3.14 units/g heart muscle}$$

1 unit = 16.67 nkat (see Appendix 1)

3.14 units = 3.14 × 16.67

$$= \textbf{52.3 nkat/g heart muscle}$$

2. (a) **Fed liver**

A   B   C   D   E   F   G

$$\frac{0.20}{0.24} \times 100 \times \frac{10}{1} \times \frac{\textbf{100}}{\textbf{1}} \times \frac{100}{1.5} \times 0.93 \times 10^{-6} \text{ g glycogen/100 g}$$

5.17 g glycogen/100 g tissue

Note: Factors in the above calculation are:

A   *ratio OD*
B   *µg glucose in standard*
C   *only 1 mL of the 10 mL hydrolysate used*
D   *only 1 mL of the 100 mL glycogen solution assayed*
E   *1.5 g tissue used, answer required for 100 g*
F   *glucose to glycogen conversion factor*
G   *to convert µg (the unit used in the calculation because the standard glucose is in µg) to g*

Fed muscle

$$\frac{0.22}{0.24} \times 100 \times \frac{10}{1} \times \frac{\textbf{10}}{\textbf{1}} \times \frac{100}{1.5} \times 0.93 \times 10^{-6} \text{ g glycogen/100 g}$$

$$= \textbf{0.57 g glycogen/100 g tissue}$$

Note: The difference in the calculations (whether the factor 100 or 10 used) is due to the volumes in which the glycogen was dissolved. The fasted tissues follow the same calculation as the fed muscle, giving **0.44 g glycogen/100 g tissue for each of fasted liver and muscle.**

(b) Performance of the Somogyi–Nelson analysis on the supernatants will show the reducing sugars (most likely glucose) in the supernatants. As no hydrolysis of glycogen is encountered here, the conversion factor 0.93 does not appear in these calculations.

Fed Liver

$$\frac{0.06}{0.24} \times 100 \times \frac{10}{1} \times \frac{100}{1.5} \times 10^{-6} \text{ g reducing sugars/100 g tissue}$$

$$= \textbf{0.017 g reducing sugars/100 g tissue}$$

(c)

	Fed		Fasted	
Table 12.7 Glycogen and Reducing Sugars in Rat Tissues				
	Liver	Muscle	Liver	Muscle
Glycogen(g/100 g tissue)	5.17	0.57	0.44	0.44
Reducing sugars(g/100 g tissue)	0.017	0.014	0.022	0.017

3. The complete calculation for the zero time mantle tissue is:

$$\overset{A}{\frac{0.37}{0.40}} \times \overset{B}{100} \times \overset{C}{\frac{10}{0.5}} \times \overset{D}{\frac{150}{1}} \times \overset{E}{\frac{93}{100}} \times \overset{F}{\frac{100}{1.3}} \times \overset{G}{10^{-6}} \text{ g glycogen/100 g tissue}$$

$$= \frac{0.37 \times 150}{1.3} \times \frac{100 \times 10 \times 93 \times 100 \times 10^{-6}}{0.40 \times 0.5 \times 100} \text{ g glycogen/100 g tissue}$$

$$= \frac{0.37 \times 150}{1.3} \times 0.465 = \textbf{19.86 g glycogen/100 g tissue}$$

Note: Repeated calculations have to be performed for each of the six sets of values. In such cases, gather the variable parts of the calculation into one set (here, the OD of the unknown, the volume of the glycogen solution, and the weight) and combine the common parts of the calculation to give a combined factor (which here works out to be 0.465). This means we have a "conversion formula" for finishing-off our "repeated calculations" (see also Answer to Question 11.4).

Factors in the complete calculation are:

A ratio OD
B µg glucose in standard (only 0.5 mL of 200 mg/mL taken)
C only 0.5 mL of the 10 mL hydrolysate used
D only 1 mL of the 150 mL glycogen solution assayed
E glucose to glycogen conversion factor
F 1.3 g tissue used, answer required for 100 g
G to convert µg (the unit used in the calculation because the standard glucose is in µg) or multiply by 10⁻⁶)

The calculations for the others, in order, are:

$$\frac{0.10 \times 150}{1.4} \times 0.465 = \textbf{4.98 g glycogen/100 g tissue}$$

$$\frac{0.32 \times 150}{1.3} \times 0.465 = \textbf{17.17 g glycogen/100 g tissue}$$

Table 12.8	Depletion of Glycogen Under Anaerobic Conditions					
	Zero Time		30 Min		1 Hour	
	Mantle	Liver	Mantle	Liver	Mantle	Liver
Glycogen per100 g tissue	19.86	4.98	17.17	2.33	17.44	0.87

$$\frac{0.15 \times 50}{1.5} \times 0.465 = \textbf{2.33 g glycogen/100 g tissue}$$

$$\frac{0.30 \times 150}{1.2} \times 0.465 = \textbf{17.44 g glycogen/100 g tissue}$$

$$\frac{0.06 \times 50}{1.6} \times 0.465 = \textbf{0.87 g glycogen/100 g tissue}$$

The glycogen values are tabulated in Table 12.8 and graphed in Figure 12.1.

4. (a) Glutamate + Pyruvate $\rightleftharpoons$ α-Ketoglutarate + Alanine K = 1.5

$$\Delta G^\circ = -RT \ln K$$
$$= -(8.3145)\,(311)\,(2.303) \log 1.5 \text{ J mol}^{-1}$$
$$= -(8.3145)\,(311)\,(2.303)\,(0.1761) \text{ J mol}^{-1}$$
$$= \textbf{-1049 J mol}^{-1}$$

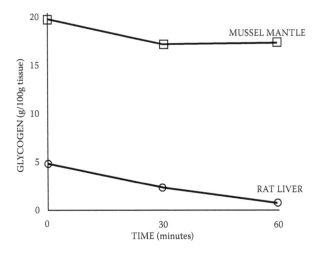

Figure 12.1   Depletion of glycogen under anaerobic conditions.

(b) Glutamate + Pyruvate $\rightleftharpoons$ α-Ketoglutarate + Alanine

$$K = \frac{[\alpha KG][Ala]}{[Glu][Pyr]}$$

Let $[Pyr] = x$ mmol/L at equilibrium

$$1.5 = \frac{(5-x)(5-x)}{(5+x)(x)}$$

$$1.5\,(5x + x^2) = 25 - 10x + x^2$$

Multiply both sides by 2 to avoid decimals:

$$15x + 3x^2 = 50 - 20x + 2x^2$$

$$x^2 + 35x = 50$$

$$x^2 + 35x - 50 = 0$$

Use formula method of solving quadratic equations:

$$x = \frac{-b \pm \sqrt{b^2 - 4ac}}{2a}$$

$$= \mathbf{1.37\ mmol/L}\ \text{or}\ -36.37$$

The negative value is ignored.

Check: (especially with this kind of mathematics!)

$$1.5 = \frac{(3.63)^2}{(6.37)(1.37)}$$

(Also check out the Internet for online sites and freewares for solving quadratic equations, if you like.)

5. Figure 12.2 is the standard curve for P_i estimation. Note that Tube 1 contains 0.4 μmol because 1 mL of 0.4 mmol/L was taken.

You can see in Figure 12.2 that the standard curve is a straight line up to OD 0.53; therefore, *it is all right* to subtract (or add) optical densities and then read the corrected OD off the curve. This will save a lot of time (and provide greater accuracy).

Firstly, each zero-time OD is subtracted from the corresponding 10-minute OD. This gives us Δ. The OD of the 'No enzyme' (0.02) is then subtracted from each of the Δ to give us the 'Corrected Δ' shown in Table 12.9. The Corrected Δ are read off the standard curve (e.g., OD 0.38 corresponds to 0.76 μmol P_i). This is the μmol P_i produced in 10 minutes (and is recorded in Column 1 of Table 12.10).

Calculation Steps

- Only 5 mL of the total 10 mL TCA-precipitated solution was assayed; therefore, for the 10 mL ($\times 2$ factor is necessary). Only 0.5 mL of each fraction was used in the enzyme assay.

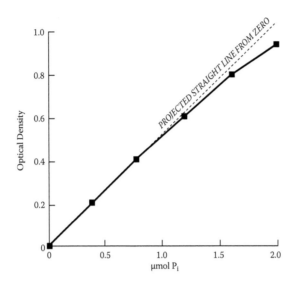

Figure 12.2 Standard curve for P_i estimation.

The Question asks for 1.0 mL, therefore another × 2 factor is involved. Finally, the enzyme incubation was for 10 minutes; units of enzyme activity requires *μmol substrate converted per 1 minute,* so divide by 10.

Overall factor is:

$$\frac{\times 2 \times 2}{10} = \times 0.4$$

(i.e., μmol P_i produced in 10 minutes [Column 1] × 0.4 to get units/mL fraction [Column 2]).

- To get values for Column 3, first multiply by 20 or 40 the mg/mL protein (given in the question, Table 12.4) to get total mg protein (Column 3). Then divide the values in Column 2 by the total mg protein to get the specific activities (Column 4).
- Enrichment in Specific Activity (Column 5) is obtained by comparing subsequent values in Column 4 to the value for the homogenate (e.g., for Nuclear Fraction):

$$\frac{0.0040}{0.0021} \times 100 = 190\%$$

Table 12.9 Corrected OD for Question 12.5		
Fraction	**Δ**	**Corrected Δ***
Homogenate	0.40	0.38
Nuclear	0.18	0.16
Mitochondrial	0.11	0.09
Microsomal	0.53	0.51
Final supernatant	0.06	0.04
No enzyme	0.02	–

Fraction	Column 1	Column 2	Column 3	Column 4	Column 5
Homogenate	0.76	0.304	144.8	0.0021	100
Nuclear	0.32	0.128	32.4	0.0040	190
Mitochondrial	0.18	0.072	31.6	0.0023	110
Microsomal	1.02	0.408	30.2	0.0135	643
Final supernatant	0.08	0.032	32.8	0.0009	43

Table 12.10   Answers for Question 12.5

Column 1 *$\mu mol\ P_i$ produced in 10 min*
Column 2 *G6Pase activity (units/mL)*
Column 3 *Total protein (mg)*
Column 4 *Specific activity (units/mg protein)*
Column 5 *% enrichment of specific activity*

(Although you were not asked to draw conclusions from the results, they appear to support what most researchers are finding — that G6Pase is a microsomal enzyme.)

6. (a) Concentration of DNA at the spectrophotometric assay level:

$$= \frac{0.302}{1.000} \times 50\ \mu g/mL$$

This solution was a 1/100 dilution because 10 μL + 990 μL. Hence, concentration of DNA in the 200 μL DNA suspension:

$$= \frac{0.302}{1.000} \times 50 \times 100\ \mu g/mL$$

In the 200 μL (or 0.2 mL), we have:

$$\frac{0.302}{1.000} \times 50 \times 100 \times \frac{0.2}{1.0}\ \mu g$$
$$= 302\ \mu g = 0.302\ mg$$

All this was contained in the original 10 mL whole blood.

$$\Rightarrow \text{concentration} = 0.302\ mg/10\ mL = \mathbf{0.030\ mg/mL}$$

(b) Ratio $\dfrac{0.302}{0.179} = 1.69$

$$\text{Percentage purity} = \frac{1.69}{1.80} \times 100 = \mathbf{94\%}$$

13

DNA and Molecular Biology

Molecular biology is the study of biology at the molecular level. In this chapter, we will deal with calculations involving deoxyribonucleic acid (DNA), ribonucleic acid (RNA), protein synthesis, and experimental methods in molecular biology.

In molecular biology calculations, there are certain average values, assumptions, or facts; these are summarised below:

1. The relative molecular masses (M_r) of amino acids range from 75 to 204. As there are more smaller amino acids in proteins, the weighted average is conveniently taken as 128 and that of amino acid residues as 110 (see answer to Question 4.10).
2. Three bases in DNA sequence are needed to code one amino acid.
3. M_r for dAMP, dCMP, dGMP, and dTMP are 331, 307, 347, and 322, respectively. They average out at 327. The average M_r for RNA bases works out as 339. (These figures are shown here only to indicate the range of values; it is not suggested that the bases appear in equal numbers in nucleic acids or oligonucleotides.) Conveniently for quick estimates and calculations, the average M_r for a DNA base is taken as 330, a base pair 660 (see Question 13.19), and RNA base as 340. (For, more accurate analysis, not only are the exact masses of each base added together but allowances are also made for alterations that occur at the ends of strands.
4. The following abbreviations are used in modern molecular biology:

   ds = double stranded
   ss = single stranded
   oligos = oligonucleotides (commonly, 2 to 10 nucleotides, but can apply to longer ones)
   -mer = from polymer or oligomer (e.g., "7-mer" means "containing 7 nucleotides").

5. Modern molecular biology techniques involve micro-, nano-, pico-, and femto- levels of analyses.
6. Melting temperatures (T_m) of dsDNA are greater when G-C base pair percentage (G-C %) is greater than A-T base pair percentage. Empirical evidence shows that T_m *varies linearly from 77° to 100° as the fraction of G-C base pairs increases from 20 to 78%.* One practical use of T_m is in the analysis of base-composition of newly isolated DNA. From the T_m value, we can get the G-C % and, in turn, the percentages of each of the individual bases. T_m determination is severely influenced by experimental conditions. One easy way around is to have standard DNA of known G-C % for comparison. Note, the relationship between T_m and base composition, quoted above, is correct in principle, but considerable variations occur due to chain lengths; nearest-neighbours, or sequence of bases; and experimental variables, such as pH and ionic concentrations. (It is considered beyond the scope of this book to include these more complicated calculations—for simpler calculations, please look at Questions 13.6 and 13.7.)

7. Absorbance at 260 nm (A_{260}) is a characteristic of all nucleotides. (For a more detailed discussion on spectrophotometry, generally please refer to Chapter 9 and specifically to the section "Protein and Nucleic Acids Analyses by UV Spectrophotometry.")

8. Spectrophotometry plays a major role in protein and nucleic acids analyses: estimating protein concentrations, estimating nucleic acids concentrations, and assessing the quality and yield of nucleic acids. Empirical values have been established as to what concentration gives an A_{260} of 1.000; these values are:

> 50 µg/mL for *ds*DNA
> 40 µg/mL for *ss*RNA
> 35 µg/mL for *ss*DNA
> 20 µg/mL for *ss* oligo DNA (particularly, this figure varies widely, as does "oligo")

These relationships are similar, in concept, to extinction coefficient or molar absorbancy (discussed in Chapter 9), except instead of trying to project what the absorbance of a molar solution would be, the baseline here is the absorbance reading of 1.000. What concentrations would give a reading of 1.000? "Mole" and "molar concentrations" are meaningless here because the molecular masses (meaning lengths of the strands) vary. These relationships and molar absorbancies are interconvertible if we know the chain length (see Question 13.14). Also, note the reading of 1.000 is, like with molar absorbancy, an "extrapolated" value from readings obtained in the lower (usually under 0.6) range.

As noted in Chapter 9, absorbance (A), rather than optical density (OD), is the preferred term for wavelengths in the ultraviolet range; although, OD_{260} is still found in recent textbooks and journal articles

Questions

1. Ribonuclease is a 124-amino acid protein. Estimate its M_r.

2. A 1.0 µmol/L solution of a commercially available *ss*DNA is serially diluted 1 in 100 twice. Calculate nmol/mL and pmol/mL for the original, the first dilution, and the second dilution.

3. A 1.0 µmol/L solution of a commercial available *ss*DNA (M_r 3300) is serially diluted 1 in 100 twice. Calculate ng/mL and pg/mL for the original, the first dilution, and the second dilution.

4. Consider an *ss*DNA 3000 bases long.
   (a) What length of polypeptide will it code?
   (b) What is the M_r of this protein?
   (c) What is the molecular mass of this protein in kDa?

5. Consider a 60 kDa protein made up of a single polypeptide. How many bases in *ss*DNA are needed to code this protein?

6. Given that T_m varies from 77° to 100° as G-C % varies from 20% to 78%, calculate the G-C % for a *ds*DNA with a T_m of 90°.

7. From the T_m value of a newly isolated *ds*DNA, it is determined to have a 60% G-C content. Calculate the percentage of each of the bases: A, G, C, and T.

8. In an experiment, a pH 7.7, 10 mmol/L Tris-1.0 mmol/L EDTA buffer is required. To allow for other additions in the setting up the experiment, the "stock" buffer needs

to be made up at 10 times the concentration. How would you prepare 200 mL of this stock?

M_r Tris 121 M_r EDTA 392

9. How would you make 200 mL 0.8% agarose in a Tris-borate-EDTA (TBE) buffer. The TBE buffer is provided (already mixed and pH adjusted) at eight times the concentration.

M_r Tris 121 M_r boric acid 62 M_r EDTA 392 M_r agarose 418

10. How would you make up 200 mL of a stock solution that is pH 7.50, 5 mol/L Tris, 50 mmol/L SDS, 1.0 mol/L NaCl, and 1% casein?

M_r Tris 121 M_r SDS (sodium dodecyl sulphate) 288 M_r NaCl 58.5

11. From 200 mL of a stock solution that is 0.5 mol/L Tris, 50 mmol/L SDS, 1.0 mol/L NaCl, and 1% casein, 0.7 mL is taken and added to 0.3 mL of a DNA solution. What is the final concentration of each ingredient in the final 1.0 mL?

12. A 1:10 dilution of a sample of *ds*DNA gave an A_{260} of 0.088. Calculate the concentration of the DNA before the dilution.

13. You have 2 µg of a 10 base pair *ds*DNA. How many moles is that?

14. Calculate the molar absorbancy:
    (a) for a 10-mer *ss*DNA.
    (b) for a 10-mer *ds*DNA

15. 0.7 mL of a stock buffer is added to 0.3 mL of a *ds*DNA solution. This mixture gave an A_{260} of 0.044. What is the µg/mL concentration of the original *ds*DNA solution?

16. To what final volume would you dilute 1.0 µL of a 36.6 mg/uL solution of DNA to get 50 ng/mL?

17. Two commonly used DNA electrophoresis buffers are Tris-acetate-EDTA (TAE) and TBE. These are available commercially (already made up) and are rated 5X, 10X, and 20X, which means they are meant to be used 1X, and appropriate dilution is allowed for while other additions may be included. Researchers can vary the concentration of these buffers further, using, for example, 0.5X instead of 1X.

    Show the amounts needed for making 200 mL agarose gel that is 0.4% agarose, 0.5X from commercial 10X TBE, and 0.5 µg/mL ethidium bromide (EtBr), a fluorescent dye that stains nucleic acids. *Caution:* It is a mutagenic hazardous chemical. To minimise handling, this too is available commercially as a 10 mg/mL solution to be (just) diluted down.

18. Consider a 9 kb (kilobase) *ds*DNA. Calculate:
    (a) the mass in mg of 1 nmol.
    (b) the number of pmol in 460 µg.

19. Calculate the M_r of an average base pair residue in DNA. You are given the relative molecular mass of free 3′-phospho deoxyribonucleosides as follows: A 331 C 307 G 347 T 322. Assume that each base in the pair is ionic bonded to two sodium atoms (A_r 23).

20. Given the average molecular weight of a base pair in DNA is 660, calculate the number of kilobase pairs in 660 g, 660 mg, 660 µg, 1 g, 1 mg, and 1 µg.

21. 50 µg/mL of *ds*DNA gives an absorbance of 1.000 at 260 nm. How many kilobases are there in 50 µg? (M_r average base pair = 660)

Answers

1. Average M_r of amino acids residues is 110.

$$\Rightarrow M_r \text{ ribonuclease} = 110 \times 124$$

$$= \textbf{13 640}$$

Note: See answer to Question 4.10 for more details.

2. Original 1 µmol/L = **1 nmol/mL = 1000 pmol/mL**

First dilution, 0.01 **nmol/mL** = 10 **pmol/mL**

Second dilution, 0.0001 **nmol/mL** = 0.1 **pmol/mL**

3. 1.0 µmol/L = 3300 µg/L = **3300 ng/mL = 3.3 × 10⁶ pg/mL**

First dilution, **33 ng/mL = 3.3 × 10⁴ pg/mL**

Second dilution, **0.33 ng/mL = 330 pg/mL**

Note: Attempt to do both Questions 13.2 and 13.3 by inspection. Cross-check your answers. Factors of 1/100 and 1/1000 are involved. Also check your answers to Question 13.3 against those for Question 13.2. The factor involved is 1/3300.

4. (a) $\dfrac{3000}{3}$ = **1000 amino acids long**

Comment: triplet code

(b) 1000 × 110 = **110 000**

Comment: M_r average amino acid = 110

(c) M_r 110 000 = **110 kDa**

Comment: M_r 1000 = 1 kDa

5. 60 kDa = M_r 60 000 = $\dfrac{60\ 000}{110}$ = 545 amino acids

$$545 \times 3 = \textbf{1635 bases}$$

Comments: Same comments as last question.

6. The range 77° to 100° or 23 degrees ≡ 20% to 78% or 58% units
Hence:

$$13 \text{ degrees} \equiv \frac{13}{23} \times 58$$

$$= 33\% \text{ units}$$

Add this to (the starting) 20% $\Rightarrow$ **53%**

Comment: This is like converting °C to °F.

7. G + C = 60%

$$G = 30\%$$

$$C = 30\%$$

$$A + T = 40\%$$

$$A = 20\%$$

$$T = 20\%$$

8. At 10x conc: **tris** 100 mmol/L   and   **EDTA** 10 mmol/L
   Per L need   100 × 121 mg   and   10 × 392 mg
   Per 200 mL   0.2 × 100 × 121   and   0.2 × 10 × 392
                    2420 mg   and   784 mg
                    2.42 g   **and** **0.784 g**

Method: Dissolve the two in about 150 mL dH$_2$O, adjust the pH on a pH meter by adding conc. HCl dropwise to pH 7.7, and then adjust the final volume to 200 mL with dH$_2$O.

Note: pK Tris is 8.14, so HCl is added to bring the pH *down* (see Table 5.2).

9. TBE is to be diluted 1 in 8, or 25 mL for 200 mL.
   Agarose at 0.8%, need 0.8 × 2 = 1.6 g for 200 mL

Method: 1.6 g agarose + 25 mL TBE → 200mL with dH$_2$O

Note: Gravimetric concentrations of agarose are involved, so the M$_r$ of agarose is not needed and neither are the other M$_r$ — do not be distracted by "red herrings" provided by lecturers or authors!

10.

	Tris 5 mol/L	SDS 50 mmol/L	NaCl 1 mol/L	Casein 1%
M$_r$	121	288	58.5	
For 1 L	× 5	× 0.05	× 1	
For 200 mL $\left(\frac{1}{5}\text{th}\right)$	121 g	2.88 g	11.7 g	2 g

Method: Add the first three compounds to about 150 mL dH$_2$O and mix. On a pH meter, adjust pH to 7.5 by adding HCl dropwise. Add casein and make up to 200 mL. (It may be best to make a paste and solubilise the casein in a little buffer before adding it to the larger volume.) *Always add compounds that are subject to denaturation, such as proteins, DNA, and the like, last.* (See answer to Question 3.22.)

Note: All above calculations can be done by simple inspection. Steps involved: taking a fifth and shifting decimal points. How do you like the table set-up for the calculations? Makes it easy for comparison and evaluation. The "1%" casein is not "qualified" as to whether w/v or w/w. In this context, it has to be w/v.

11. The dilution of the original concentration is $\dfrac{0.7}{1.0}$. So, multiply each by 0.7:

$$0.5 \text{ mol/L Tris} = \mathbf{0.35 \text{ mol/L}}$$

$$50 \text{ mmol/L SDS} = \mathbf{35 \text{ mmol/L}}$$

$$1.0 \text{ mol/L NaCl} = \mathbf{0.7 \text{ mol/L}}$$

$$1\% \text{ casein} = \mathbf{0.7\%}$$

12. Absorbance 1.000 given by 50 µg/mL

$$\Rightarrow 0.088 \text{ given by } 0.088 \times 50 \text{ µg/mL}$$

Before dilution:

$$0.088 \times 50 \times 10 \text{ µg/mL}$$

$$= \mathbf{44 \text{ µg/mL}}$$

13. 10 base pairs $= 660 \times 10 = 6600 \text{ M}_r$

$$2 \text{ µg} = \frac{2}{6600} = 3.03 \times 10^{-4} \text{ µmol}$$

$$= 303 \times 10^{-6} \text{ µmol}^*$$

$$= \mathbf{303 \text{ pmol}}$$

*__Note:__ See answer to Question 2.17 for advice on how to handle this kind of expression.

14. (a) 40 µg/mL *ss*DNA gives an A_{260} of 1.000.

$$\Rightarrow 330 \times 10 = 3300 \text{ µg/mL gives an } A_{260} \text{ of } \frac{3300}{40} \times 1.000$$

$$= 82.5$$

And 3300 mg/mL (1 molar) would give $82.5 \times 10^3 = \mathbf{8.25 \times 10^4}$

__Note:__ Instead of the ratio method used above, the formula given in Chapter 9 may be used.

$$\text{Concentration (mol/L)} = \frac{\text{Absorbance}}{\varepsilon \times \text{pathlength (cm)}}$$

"40 µg/mL *ss*DNA" means 40 µg/mL of *any* M_r *ss*DNA.

For our 10-mer *ss*DNA, the mol/L value is $\dfrac{40}{3300}$ µmol/mL.

$$= \frac{40}{3300} \text{ mmol/L}$$

$$= \frac{40}{3300} \times 10^{-3} \text{ mol/L}$$

Apply this value in the formula:

$$\varepsilon = \frac{1}{\dfrac{40 \times 10^{-3}}{3300}} = \frac{3300}{40 \times 10^{-3}} = 8.25 \times 10^4$$

Note: Often, more mistakes are made in converting small concentrations to mol/L than at other steps. It is recommended that you do this conversion by inspection rather than by applying another formula. The final answer is expected to be a big number—in thousands or tens of thousands; after all, it is supposed to be the reading of a 1 mol/L solution! (Remember the ε for NADH?) If you get an answer 10^{-3} or 10^{-4}, better start again.

(b) The M_r for the 10-mer dsDNA is 6600. 50 µg/mL gives a reading of 1.000. Following the formula method from above:

$$\varepsilon = \frac{6600}{50 \times 10^{-3}} = 1.32 \times 10^5$$

Note: Let us try to evaluate this answer. It is 1.6 times that of (a) — not quite double. Suppose 40 µg/mL dsDNA had given the reading 1.000; then substituting 40 in place of 50 (in the last equation above) would give an answer 1.65×10^5, which is exactly double. Due to chemical reasons, we needed all of 50 µg of dsDNA to give the same reading as 40 µg ssDNA. Since 50 was used in the denominator, we got an answer which is not quite double the answer for (a). The chemical reasons referred to above are not our concern now, but you may recall seeing that A_{260} increases in DNA melting curves when dsDNA becomes ssDNA. It would be worthwhile remembering that A_{260} of ssDNA is about 20% greater than that for the same amount of dsDNA.

15. An A_{260} of 1.000 is given by 50 µg/mL dsDNA solution.

$$\Rightarrow 0.044 \text{ is given by } \frac{0.044}{1.000} \times 50 \text{ µg/mL}$$

At the assay level, the DNA solution was diluted $\dfrac{0.3}{1.0}$.

$$\Rightarrow \text{ original DNA solution is } \frac{1.0}{0.3} \times \frac{0.044}{1.000} \times 50 \text{ µg/mL}$$

$$= 7.33 \text{ µg/mL}$$

16. The concentration ratio is, 36 600 ng/mL:50 ng/mL

$$\text{or } \frac{36600}{50} : 1$$

$$= 732 : 1$$

So dilution ratio must be the other way about.[*]

Dilute 1.0 µL to 732 µL.

[*] **Note:** If you do not follow this form of reasoning, please review the section on ratio dilutions in Chapter 3.

17. Note, all concentrations are in gravimetric units. For 200 mL:

Agarose 0.8 g

TBE 0.5X from 10X

$$\Rightarrow 1 \text{ in } 20, \text{ or } \mathbf{10 \ mL} \text{ in } 200 \text{ mL}$$

EtBr is 10 mg/mL, or 10 000 µg/mL.
Need 0.5 µg/mL

$$\Rightarrow \text{ for } 200 \text{ mL, need } 100 \text{ µg } (yes?)$$

$$(\text{i.e., } 0.01 \text{ mL or } 10 \text{ µL})$$

Note: Expect tiny amount of EtBr; it is only a dye.

18. (a) M_r of a 9 kb dsDNA $= 9000 \times 660$

$$= 5.94 \times 10^6$$

$$1 \text{ nmol} = 5.94 \times 10^6 \text{ ng}$$

$$= \mathbf{5.94 \ mg}$$

(b) $460 \text{ µg} = \dfrac{460}{5.94 \times 10^6} \text{ µmol}$

$$= 7.74 \times 10^{-5} \text{ µmol}$$

$$= \mathbf{77.4 \ pmol}$$

19. A-T pair $= 331 + 322 - 2(18) = 617$

$$\text{G-C pair} = 347 + 307 - 2(18) = 618$$

$$\text{Average} = \dfrac{617 + 618}{2} = 617$$

$$\text{Add two Na} = 617 + 2(23) = \mathbf{663}$$

Note: A pairs with T, and G with C; each nucleotide loses one molecule of water when linked to form a polynucleotide. It is assumed there are equal numbers of A-T and G-C pairs in the DNA in this question; even without this assumption, the answer would be practically the same. Why? When detached from their histones, these polyanions appear as salts commonly with sodium.

20. 660 g is the mass of 6.02×10^{23} base pairs (Avogadro's No.)

$$= 6.02 \times 10^{20} \text{ kilobase pairs}^*$$

$$660 \text{ mg} = 6.02 \times 10^{17} \text{ kb pairs}$$

$$660 \text{ µg} = 6.02 \times 10^{14} \text{ kb pairs}$$

1 g is the mass of $\dfrac{1}{660}$ × 6.02 × 10²³ base pairs

$$= 9.12 \times 10^{20} \text{ base pairs}$$

$$\mathbf{= 9.12 \times 10^{17} \ kb \ pairs^{*}}$$

$$1 \text{ mg} = \mathbf{9.12 \times 10^{14} \ kb \ pairs}$$

$$1 \text{ μg} = \mathbf{9.76 \times 10^{11} \ kb \ pairs}$$

Check: 660:1 should be the same as 6.02 × 10²⁰ kilobase pairs* : 9.12 × 10¹⁷ kb pairs*.

Yes, it works out.

21. 660 g is the mass of 6.02 × 10²³ base pairs (Avogadro's No.)

50 μg in the mass of $\dfrac{50 \times 10^{-6}}{660}$ × 6.02 × 10²³

$$= 4.56 \times 10^{16} \text{ base pairs}$$

$$\mathbf{= 4.56 \times 10^{13} \ kilobase \ pairs}$$

14

Pharmaceutical Calculations

Pharmaceutical calculations are not difficult. In the preface to the first edition of this book, it was maintained that to successfully employ the *ratio method,* "no advanced knowledge of mathematics is required beyond the elementary logic behind mathematical operations acquired in junior secondary school." This is certainly true for performing pharmaceutical calculations. Having said that, the pharmacy student should equally take notice of the fact that pharmaceutical calculations are *not* easy. In junior high school, you were not expected to get your calculations correct 100% of the time. In pharmaceutical calculations, there is no margin for error whatsoever.

Although pharmaceutical calculations may be easy, a whole set of rules are expected to be followed by the pharmacy student (as well as the practitioner). While students may think this is all so boring and "I can easily handle this in my head," please take note of what pharmacy education authorities have advocated for years. Heed the warning: Get it absolutely right *all* the time, and further, make certain that your numerical values are presented in such a manner that nobody else will misinterpret them! Your future patients' lives depend on it.

Using proportions to do calculations (the *ratio method,* though not necessarily called by this name) is the preferred method for doing pharmaceutical calculations.

When you are required to make up a prescription, it is strongly advised that you slow down your work considerably. Do not string blocks together in one line (as suggested earlier in this book); instead, do each step separately. The added bonus of this approach is the opportunity to evaluate each step. Remember not to round-off too severely, as this can introduce round-off errors.

Two important rules must be followed in the presentation of your numerical values:

1. **Numerical values should *never* contain a naked decimal point.** This is an example of a number containing a naked decimal point: .1. Students and practitioners should never write numbers this way. It can be mistaken for 1 and, when dealing with medical dosages, it would mean a 10-times stronger dosage! Always "protect" the decimal point with a zero. Always write 0.1, 0.227, and so forth.
2. **Numerical values should *never* contain a trailing zero.** This is an example of a number with a trailing zero: 1.0. "Now what's wrong with that?" students may ask and equally challenge the suggested alternative, 1, arguing that there is a difference between 1 and 1.0. Science students will correctly claim that the zeros after the decimal point convey precision, as in "number of significant figures." However, pharmacy teachers argue that 1.0 can be (and has been) mistaken for 10, with disastrous consequences.

Pharmacy students are expected to follow these two rules of writing numbers right from first year and continue that practice as practitioners.

Pharmaceutical calculations are otherwise not unlike those encountered in the earlier chapters of this book. There is, however, an additional dimension that is introduced in this patient-care profession—that is, patient weight is not expressed in scientific (let alone SI) units in many countries. Further, elderly patients, particularly, may be more comfortable with Imperial units. In all interactions with patients, or laypersons in general, the scientist is expected to be charitable (*and learned*) to convert information into the "language" that the public can understand. (Please look at Appendix 1, Units and Their Conversions.)

Checking Answers

1. As practicing pharmacists, it is likely that you may have other pharmacists around you. Have one of your colleagues check your answer. It is best that they do it entirely independently.
2. If you are by yourself, verify the answer by taking a totally different approach to the calculation. First, perform the calculation by methods suggested in this book, using only an electronic calculator. Then, if you may have software or your own (familiar and personally created and tested) formula on your computer, plug the figures and see what comes out.
3. Also, if you are on a good thing, stick to it. Keep records of all your calculations. If there is a similar problem, use the same approach and compare the new answer to the old one.
4. Estimate the answer. In multistep calculations, it should be quite easy to estimate each step as you go along. Even thinking "should be about half" would keep you on the right track. Especially with conversions, "would expect a smaller (or larger) number" would confirm that you have done the conversion the right way. Estimation is a very powerful tool. In all cases, it provides an approximation as to where the answer should fall. Furthermore, estimation is only done by using your gray matter, and that is an excellent thing! Keep your senses. If you are tuned in, you will pick up that *a 10-lb baby cannot weigh 22 g.*
5. Finally, ask yourself: Does this answer make sense? All people make mistakes. With all the best intentions and checking procedures, you could have copied something wrong or missed writing down 10^3 (or written it as 103). Be happy if you recognise a nonsensible answer — and you can go back and correct it.
6. Writing 981 instead of 891 will not possibly show up in the final does-it-make-sense checking? Let us hope that writing 912 instead of 192 does. If you are prone to dyscalculia, which is a form of dyslexia (do not worry: it is very common and occurs across the IQ range), you are obviously aware of it. If you have ever transposed numbers (we have all done that at some stage), make a point of going back and checking your figure entries. Allow yourself one check-through, checking figure entries only.
7. Do anything — and everything — so that the so-called "disastrous consequences" never happen to your patient and you!

Questions

(For questions involving pharmaceutical preparations, set out the steps of your calculations separately. Do not string out the blocks of ratios.)

1. 0.1234 g of compound C was dissolved in water to produce 1000 mL final volume. How many micrograms of compound C are contained in 1 mL of this solution?

2. A preparation of a pharmaceutical compound is 3.5 g/100 mL. 10 mL of the preparation was made up to 500 mL with water, then 20 mL of the resulting solution was made up to 100 mL with water, and finally, 5 mL of the 100 mL was made up to 500 mL with water. How many μg of the pharmaceutical compound are there in 5 mL of the final solution?

3. A pharmaceutical compound (Rx) is available as a concentrated (water soluble) paste. 0.1406 g of the paste was made up to 100 mL with water. This "original" solution was diluted serially *twice further* (each time a 100-fold [i.e., 1 mL to 100 mL]). The final dilution was analysed and found to contain 36.68 ng Rx/mL. What was the concentration (% w/w) of the paste?

4. Two brands of eyedrops designed to "clear red eye" are sold in 15 mL plastic bottles. The active ingredient in both is naphazoline hydrochloride. Brand A is quoted as 1 mg/mL and sells for $6.99, and Brand B as 0.3% w/v for $5.99. Based only in the amount of naphazoline hydrochloride, which is a better value?

5. A suspension contains 400 mg of amoxicillin per 5 mL.
   (a) If the patient's dose is 150 mg, what volume should the patient receive?
   (b) Naturally, it would be easier for the patient to take her dosage as a teaspoon-full (5 mL) rather than some irregular volume like 1.76 mL. And, if the patient were required to take the dose twice-a-day for 7 days, how would you prepare the total volume required using the commercial fluid AmoxDiluent?

6. A suspension contains 400 mg of amoxicillin per 5 mL. What amount is contained in 18.8 mL?

7. A child who weighs 37 lbs is prescribed 5 mg/kg/day of diphenhydramine HCl to be taken "4 times a day." The diphenhydramine HCl is available as syrup, whose diphenhydramine HCl concentration is 12.5 mg/5 mL. Assuming that "4 times a day" means equal doses every 6 hours, what should be the volume of syrup of each dose?

8. Suppose a compound PRS is available as a 30% (w/w) paste and is prescribed to be made up as 2 mg/mL in 40% (v/v) isopropyl alcohol and 100 mL is required. The compounding pharmacist's source of isopropyl alcohol is a 70% (v/v) solution. Show how this preparation should be made up.

Answers

1. 0.1234 g/L = 123.4 mg/L = **123.4 μg/mL**

   Comment: Do conversions by inspection.

2. $3.5 \times \dfrac{10}{100} \times \dfrac{20}{500} \times \dfrac{5}{100} \times \dfrac{5}{500} \times 10^6 = \mathbf{7\ μg}$

   Note: Dilution factors are not needed. Please make certain that you understand how each multiplicand ratio is created and you feel confident to "string out the blocks."
   (If you find the above "stringing out" too winding, consider breaking it up as shown in the alternative approach below.)

Comments:

The thinking goes like this:
There 100 mL to start with.
Of this, 10 mL was taken.
This is now in the 500 mL.
Of this, 20 mL was taken.
This is now in 100 mL.
Of this, 5 mL was taken.
This is now...

Note: All the multiplicand ratios are "fractions smaller than one." They only represent that small portion that was actually used—the remainder discarded. So, we end up with a very small answer: 0.000 007g. But by calling this number in µg (the name given to tiny fractions of grams), we create a big number (of micrograms). To do this, we must multiply by 10^6 (not, 10^{-6}!). These are all matters we should be aware of—and consciously do.

Alternative Approach
0.35 g in 500 mL
20 mL taken

i.e., $0.35 \times \dfrac{20}{500}$ g

This in 100 mL, of which 5 mL taken

i.e., $0.35 \times \dfrac{20}{500} \times \dfrac{5}{100}$ g

This in 500 mL, of which 5 mL take

[above expression] $\times \dfrac{5}{100}$

$= 0.\,000\,007$ g

$= \textbf{7 µg}$

Comment: The first step is not written down. By inspection, we realise that in taking 10 mL of a 3.5 g/100 mL solution, we are in fact starting off with 0.35 g. (Sometimes things are "done in the head" to avoid clutter. Use a pencil or pen to point and always double-check.)

3. Each dilution 10^2, overall 10^4
   36.68×10^{-9} g/mL becomes 36.68×10^{-5} g/mL
   or 36.68×10^{-3} g/ 100 mL 'original'

   Compare 36.68×10^{-3} g to 0.1468 g $= \dfrac{0.03668}{0.1406} \times 100$

   $$= \textbf{26\%}$$

4. There are alterative approaches to this kind of question. Take the simplest approach: In this case, covert both to g/mL.
   A 1 mg/mL = 0.001 g/mL
   B 0.3 g/100mL = 0.003 g/mL
   $\Rightarrow$ **B** is a better value. With pharmaceutical preparations, however, other factors must be considered, such as other additives, ease of use, and so forth. These points are not considered here.

5. (a) 400 mg in 5 mL

$$\Rightarrow 150 \text{ mg in } \frac{150}{400} \times 5 \text{ mL}$$
$$= \mathbf{1.88 \ mL}$$

Check: The answer should be less than half of 5 mL.

5. (b) Total volume required = 70 mL (i.e., 14 doses)
Each dose is 1.88 mL (calculated in part (a))
14 doses = 1.88 × 14 = **26.32 mL concentrate**
Dilute this to 100 mL.

6. 5 mL contains 400 mg.

$$18.8 \text{ mL contains } \frac{18.8}{5} \times 400 \text{ mg}$$
$$= 1504 \text{ mg}$$
$$= \mathbf{1.50 \ g}$$

Note: Work in mg. The answer should be more than 3 times 400. Convert milligrams to grams by inspection. Do not perform an operation dividing by 1000 or worse, 10^3, leaving yourself open for errors that might creep in. Check the conversion, though — no matter how simple it is.

7. **Step 1.** Convert pounds to kilograms.

$$37 \times 0.4536 = 16.78 \text{ kg}$$

(Conversion factor given in Appendix 1.)
Step 2. 1 kg requires 5 mg

$$\Rightarrow 16.78 \text{ kg requires } \frac{16.78}{1} \times 5 \text{ mg}$$
$$= 83.9 \text{ mg/day}$$

Comments: If you feel comfortable with 5 × 16.78 (and you are absolutely sure), do that. There is no need to set up the 16.78 over 1.

Check: 16 × 5 = 80, so 83.9 should be right.

Step 3. Divide into four doses.

$$\frac{83.9}{4} = 20.98 \text{ mg/day}$$

Check: It is a quarter of about 80, so it is about 20.

Step 4. In what volume of solution?

$$\Rightarrow 12.5 \text{ mg in 5 mL}$$
$$\Rightarrow 20.98 \text{ mg in } \frac{20.98}{12.5} \times 5 \text{ mL}$$
$$= 8.392 \text{ mL}$$
$$= \mathbf{8.4 \ mL}$$

Check: It is not quite double 5 mL.
(While it is not required in the question, instead of ending with an answer of 8.4 mL, further dilution as in Question 5(b) could make the volume 10 mL, allowing the dose to be two teaspoonfuls.)

8. **Step 1.** PRS is to be 2 mg/mL.
Want 100 mL

$$\Rightarrow 100 \times 2 \text{ mg} = 200 \text{ mg required}$$

Step 2. As the paste is not pure, but only 30%, we need:

$$200 \times \frac{100}{30} = 666.66 \text{ mg}$$

Check: A larger mass by the ratio of $\frac{100}{30}$ is required

$$666.66 \text{ mg} = \textbf{0.666 g}$$

Check: Are you sure? Always do these conversions by inspection.

Step 3. Isopropyl is 70%, and we need to get it down to 40%.

$$70 \text{ mL in } 100 \text{ mL}$$
$$40 \text{ mL in } \frac{40}{70} \times 100 \text{ mL}$$

$$= \textbf{57.14 mL}$$

(This is eventually to be diluted to 100 mL.)

Note: Make these kinds of action statements: "Get it *down to* 40%," so we are *only* taking 57.14 mL which, when diluted to 100 mL, will get 70% *down to* 40%.

Step 4. (putting it all together)
Weigh out 0.667 g of the paste. Measure out 57.14 mL 70% isopropyl alcohol as accurately as possible in a 100 ml graduated cylinder. Wash out the entire paste with the isopropyl alcohol into a 100 mL or 200 mL beaker. Quantitatively transfer this to a 100 mL standard flask and make the volume to 100mL with distilled water.

Check: Keep close check of the materials: whether it is solute (paste) or solvent (isopropyl alcohol). Clearly identify which you are dealing with at each step (refer to them by name). There are two dilutions taking place here, each with its own specifications; don't confuse one with the other.

Comment (on procedure)
Judging from the information given in the question, the paste is not water-soluble, hence, the compounding pharmacist should not expose the paste to too much water. The approach taken here is to dissolve the paste in 70% isopropyl alcohol before diluting it (the isopropyl alcohol) to 40%.

Appendix 1
Units and Their Conversions

The SI (system of units) is the officially approved unit system used by the international science fraternity. However, as partly discussed in Chapter 2, there are other units besides (or officially called "outside") the SI that are used in science and even permitted by top international scientific journals. There are also "measures" used by the nonscientific world ("commercially" or by the "layperson"). In this appendix, an attempt is made to (a) list many of the units that are mainly used by biological and chemical scientists and (b) show the conversions to other units or measures.

As much as possible, the comments and recommendations that are included try to match the guidelines of the British *Biochemical Journal (BJ)* and/or the U.S. *Journal of Biological Chemistry (JBC)*.

Table A1 SI Base Units		
Base Quantity	**Name**	**Symbol**
length	metre	m
mass	kilogram	kg
amount of substance	mole	mol
time	second	s
thermodynamic temperature	kelvin	K

Table A2 SI-Derived Units		
Derived Quantity	**Name**	**Symbol (Expression)**
area	square metre	m^2
volume	cubic metre	m^3
concentration (of amount of substance)	mole per cubic metre	mol/m^3
force	newton	$N \ (m \ kg \ s^{-2})$
pressure	pascal	$Pa \ (N/m^2)$
temperature (scientific)	degree Celsius[a]	$°C \ (K + 273.15)$
energy, work, quantity of heat	joule	$J \ N \cdot m \ (= m^2 \ kg \ s^{-2})$

[a]While not in the official list, it is completely acceptable and convenient to include here. Intervals in K and °C are the same.

Table A3   Accepted[a] Units Outside SI

Name	Symbol	Value in SI
minute	min	1 min = 60 s
hour	h	1 h = 60 min
day	d	24 h
litre[b]	L (or l)	1 L = 1 dm^3
metric ton[c]	t	1 t = 10^3 kg
unified atomic mass unit	u	1 u = 1.660 54 × 10^{-27} kg, approximately

[a] These have been accepted for use alongside the SI units by the International Committee for Weights and Measures.

[b] Both symbols, L and l, are internationally acceptable. L is preferred (especially in Australia and the United States) as it avoids confusion with 1 (one) and the letter I.

[c] Many countries that have gone metric use the metric ton, spelling it "tonne" (to avoid confusion with the old Imperial ton).

Table A4   SI-Approved Prefixes for Units

Factor	Prefix	Symbol
$10^{24} = (10^3)^8$	yotta	Y
$10^{21} = (10^3)^7$	zeta	Z
$10^{18} = (10^3)^6$	exa	E
$10^{15} = (10^3)^5$	peta	P
$10^{12} = (10^3)^4$	tera	T
$10^9 = (10^3)^3$	giga	G
$10^6 = (10^3)^2$	mega	M
$10^3 = (10^3)^1$	kilo	k
10^2	hecto[b]	h
10^a	deka[b]	da
10^{-1}	deci[b]	d
10^{-2}	centi[b]	c
$10^{-3} = (10^3)^{-1}$	milli	m
$10^{-6} = (10^3)^{-2}$	micro	μ
$10^{-9} = (10^3)^{-3}$	nano	n
$10^{-12} = (10^3)^{-4}$	pico	p
$10^{-15} = (10^3)^{-5}$	femto	f
$10^{-18} = (10^3)^{-6}$	atto	a
$10^{-21} = (10^3)^{-7}$	zepto	z
$10^{-24} = (10^3)^{-8}$	yocto	y

[a] In SI language, the term "submultiple" is used to mean "fraction" like those for deci and smaller.

[b] The current preference is to avoid using hecto, deka, deci, and centi, except for centi in centimetre.

Table A5   Temporarily Accepted Units Outside the SI

Name	Symbol	Value in SI Units
Nautical mile		1 nautical mile = 1852 m
Knot		1 nautical mile per hour = (1852/3600) m/s
Are[a]		100 m^2
Hectare[b]	ha	= 100 are = 1 hm^2 = $10^4\,m^2$
Ångström[c]	Å	1 Å = 0.1 nm = 10^{-10} m

[a] This unit, pronounced "air," is not commonly used.
[b] A multiple of are, the hectare (ha) is used for land acreage measurements.
[c] An older unit used for designating wavelength. Nanometre (nm) is now the preferred unit.

Table A6   Commonly Used Molecular Units and Other Terms

Unit	Symbol	Value/Comments
Mole	mol	SI unit for amount of chemical substance M_r expressed in grams Contains as many elementary entities as 12g ^{12}C (i.e., Avogadro's No. 6.022×10^{23}) Mole is also used with atoms; it is relative atomic mass (A_r) in grams
Avogadro's number	N_A	Number of elementary entities in 1 mole of any substance incl. 12g ^{12}C
Relative molecular mass	M_r	A relative "molecular mass" unit that now replaces the older molecular weight (MW) and even-older formula weight (FW). The ratio of the mass of the molecule to $\frac{1}{12}$ of the mass of carbon^{12} and is dimensionless
Relative atomic mass	A_r	Comparable to M_r, is the term replacing atomic weight (AW)
Unified atomic mass unit	uamu (old)	Very small unit of mass; mass of $\frac{1}{12}$ of one atom of ^{12}C = 1/Avogadro's No. 6.022×10^{23} g = 1.661×10^{-24} g
Dalton	Da	Unit of mass = u, used with proteins/macromolecules, bioaggregates. 1 kilodalton kDa = 10^3 Da. Favoured by biologists
Molecular mass	m	Actual mass of a molecule in Da or u
Enzyme unit	U	Amount of enzyme that catalyzes the conversion of 1 micro-mole of substance per minute. 1U = 16.67 nanokatal
Katal	kat	The amount of enzyme that converts 1 mole of substrate per second. 1 kat = 6×10^7 U. Favoured over U because it is SI related. Katal is excessively large, so subunits, such as microkatal and nanokatal, are used
International unit	IU	Unit for measuring biological activity of vitamins, hormones, drugs, and such. Definition parameters vary for different substances. Not related to enzyme unit

Table A7	Commonly Used Gravimetric Concentration Units
Name	**Comments/Value**
g/100 mL	g per 100 mL of final solution
g/L	g per litre of final solution
mg/mL	= g/L
μg/mL	= 10^{-6} g/L
ng/mL	= 10^{-9} g/L
% (w/v)	g/100 mL final solution
% (w/w)	g/100 g of final solution or paste
% (v/v)	mL/100 mL final solution
ppm	(parts per million) = μg/mL = mg/L liquids, or = μg/g = g/tonne solids (1ppm = 0.0001% w/v or w/w)

Table A8	Commonly Used (and Preferred) Chemical Concentration Units	
Name	**Preferred Symbol**	**Former Symbol**
Molar	mol/L	M
Millimolar	mmol/L	mM
Micromolar	μmol/L	μM
Nanomolar	nmol/L	nM
Picomolar	pmol/L	pM
Femtomolar	fmol/L	fM

Table A9	Some Common U.S. Customary Measures of Length and Conversions
Length Unit	**Equivalent**
1 inch	2.54 cm
1 foot (12 inches)	30.48 cm
1 yard (3 feet)	0.9144 m
1 mm	0.03937 inch
1 cm	0.3937 inch
1 m	1.09361 yard, 3.281 ft

Table A10   Some Common U.S. Customary Measures of Mass and Conversions

Mass Unit	Equivalent
1 dram	1.77184 g
1 ounce (16 drams)	28.35 g
1 pound (16 ounces)	0.4536 kg
1 stone[a] (14 pounds)	6.3492 kg
1 ton[a] (2240 pounds)	1 016 kg
1 short ton (2000 pounds)	907.2 kg
1 gram	0.56438 dram = 0.353 oz
1 kilogram	2.205 pounds

[a] *Not part of U.S. customary measures*

Table A11   Some Common U.S. Customary Measures of Volume and Conversions

Volume Unit	Equivalent
1 fluid dram	3.70 mL
1 fluid ounce (8 fluid drams)	29.57 mL
1 liquid pint (16 fluid ounces)	0.4732 L
1 liquid quart (2 liquid pint)	0.9464 L
1 liquid gallon (4 liquid quart)	3.7854 L
1 mL	0.027 fl dr = 0.034 fl oz
1 L	2.1134 pints = 1.057 quarts = 0.2642 gallons

Table A12   Miscellaneous Units, Conversions, and Comments

Unit	Symbol	Equivalent
1 joule	J	10^7 ergs[a]
1 kilojoule	kJ	1000 J
1 calorie (food/nutrition)	cal	4.184 J
1 calorie[b] (food/nutrition)	Cal (or kcal)	1000 cal = 4.184 kJ
1 calorie[c] (heat/engineering)	cal	4.186 J
Fahrenheit degrees[d]	°F	$°F = 1.8 (°C) + 32$ $$°C = \frac{(°F - 32)}{1.8}$$

[a] Historically, erg is a tiny fraction of a joule; now submultiples, such as μJ, are preferred.
[b] Also known as a "large" calorie or kilocalorie. Confusing and is best avoided. Use J or kJ instead.
[c] Another reason to avoid "calorie"—large or small.
[d] Of nonscientific use in the United States; almost everywhere else °C (Celsius, formerly Centigrade).

Table A13   Approximate Equivalents of Liquid Volumes

Utensil	Approximate Metric Volume
1 teaspoonful	5 mL
1 tablespoon = 3 teaspoonfuls	15 mL
1 cup	250 mL
2 cups = 1 liquid pint	500 mL
1 household bucket	10 L
44 (imperial) gallon drum = 55 (U.S.) gallon drum	210 L

Suggested Reading and Bibliography

Dawson RMC, Elliott DC, Elliott WH and Jones KM (1990) *Data for Biochemical Research,* 3rd ed. Oxford, New York. Unlike most data books, this book is a pleasure to look through. It is totally relevant to biochemistry and contains factual information on compounds, reagents, and techniques—a must-have for any biomedical research laboratory bookshelf and good reference book for undergraduates.

Dennison C (1988) *A Simple and Universal Method for Making up Buffer Solutions.* Biochemical Education 16(4) p 210-211. This paper discusses and advocates the calculation-titration method (see Chapter 5 of this book) for the preparation of buffers, including buffers of defined ionic strength.

Holme DJ and Peck H (1998) *Analytical Biochemistry,* 3rd ed. Longman, London. Directed at undergraduates, this book covers all major biochemical analytical techniques and discusses principles and background information, allowing students to compare different techniques; information on data handling and writing reports is included.

Holum JR (1990) *Fundamentals of General, Organic, and Biological Chemistry,* 4th ed. Wiley, New York. This book is recommended to students who feel they have missed out on basic chemistry. It includes a large number of elementary quantitative problems—fully worked-out examples and practice exercises with final answers. It also includes an appendix on mathematical concepts suitable for chemical and biochemical calculations.

Jack, RC (1995) *Basic Biochemical Laboratory Procedures and Computing: With Principles, Review Questions, Worked Examples, and Spreadsheet Solutions.* Oxford, New York. This title, which is self-explanatory, is better suited for advanced undergraduates or research students.

Lehninger AL, Nelson DL and Cox NM (2004) *Principles of Biochemistry,* 4th ed. Freeman, New York. A popular undergraduate textbook for the last three decades, this book is extremely readable. It contains a limited number of quantitative questions with final answers. An accompanying study guide, *The Absolute, Ultimate Guide* by Osgood & Ocorr, is available.

Montgomery R and Swenson CA (1976) *Quantitative Problems in the Biochemical Sciences,* 2nd ed. Freeman, San Francisco. This book contains a fair number of elementary chemical calculations as well as the standard undergraduate biochemistry questions. It also includes theory, worked exercises, and practice questions. A useful text, unfortunately no later edition is available and the book is now out of print.

Plummer DT (1987) *An Introduction to Practical Biochemistry,* 3rd ed. McGraw-Hill, London. This text contains good descriptions of many undergraduate biochemical techniques, theories, and principles related to quantitative experiments and a limited number of calculations with final answers.

Roger LL (2007) *Biochemistry and Molecular Biology Compendium* CRC Press, Taylor & Francis, London. This compendium provides information relating to biochemistry, molecular and systems biology, biotechnology, proteomics, and genomics not found in more database-oriented resources. The information is targeted at researchers; advanced undergraduate will find the glossary of terms and acronyms quantitative data both useful and educational.

Segel IH (1976) *Biochemical Calculations: How to Solve Mathematical Problems in General Biochemistry,* 2nd ed. Wiley, New York. This book contains relevant theory, fully worked-out solutions with advisory comments, and a large number of practice questions with final answers. Although this book is still prescribed, the 1976 edition is the latest available.

Stryer L (1995) *Biochemistry,* 4th ed. Freeman, New York. This book enjoys similar popularity to Lehninger, especially in more recent years. It contains a limited number of quantitative questions with final answers. A *Student's Companion* to Stryer's *Biochemistry* by Gumport, RI, et al., is also available.

Wilson K and Goulding KH (1993) *Principles and Techniques of Practical Biochemistry,* 4th ed. Hodder, London. This book provides excellent readable accounts of techniques commonly used in biochemistry and molecular and cell biology. It also offers comprehensive coverage of principle and theory and includes tables comparing techniques.

Yates P (2007) *Chemical Calculations: Mathematics for Chemistry,* 2nd ed. CRC Press, Taylor & Francis, London. This book provides a unified reference of mathematical techniques in the context of chemistry, thus providing an ideal reference for undergraduate *chemistry* students who may have missed out on solid mathematical foundations in the latter years of high school; numerous worked examples along with exercises and soluations are included.

Index

9 781420 053579